Raising Robots to be Good

Rebecca Raper

Raising Robots to be Good

A Practical Foray into the Art and Science of Machine Ethics

Rebecca Raper
Centre for Robotics and Assembly
Cranfield University
Coventry, UK

ISBN 978-3-031-75035-9   ISBN 978-3-031-75036-6 (eBook)
https://doi.org/10.1007/978-3-031-75036-6

© The Editor(s) (if applicable) and The Author(s), under exclusive license to Springer Nature Switzerland AG 2024

This work is subject to copyright. All rights are solely and exclusively licensed by the Publisher, whether the whole or part of the material is concerned, specifically the rights of translation, reprinting, reuse of illustrations, recitation, broadcasting, reproduction on microfilms or in any other physical way, and transmission or information storage and retrieval, electronic adaptation, computer software, or by similar or dissimilar methodology now known or hereafter developed.
The use of general descriptive names, registered names, trademarks, service marks, etc. in this publication does not imply, even in the absence of a specific statement, that such names are exempt from the relevant protective laws and regulations and therefore free for general use.
The publisher, the authors and the editors are safe to assume that the advice and information in this book are believed to be true and accurate at the date of publication. Neither the publisher nor the authors or the editors give a warranty, expressed or implied, with respect to the material contained herein or for any errors or omissions that may have been made. The publisher remains neutral with regard to jurisdictional claims in published maps and institutional affiliations.

This Springer imprint is published by the registered company Springer Nature Switzerland AG
The registered company address is: Gewerbestrasse 11, 6330 Cham, Switzerland

If disposing of this product, please recycle the paper.

Never let your sense of morals prevent you from doing what is right.
– Isaac Asimov (1960)

Everything's got a moral, if only you can find it.
– Alice from 'Alice's Adventures in Wonderland' (Lewis Carroll, 1865)

Inspired by and made for Lilly and Jack.

Preface and Acknowledgements

Though the outcome of my PhD research, this book has been the culmination of 7 years intense thinking around how we might create machines with morals. I was first introduced to the topic back in 2017 when I saw a PhD advertisement to work on the topic of 'Robots and Kindness'. Up until this point I had only studied philosophy (I had a particular interest in *logic* and *the philosophy of mind*) and a bit of psychology, and knew I wanted to apply my experience to solving problems in Artificial Intelligence but didn't know how. When I saw the project scope asking for a philosopher to look at how we might create *kind robots*, I thought the project was ambitious and crazy enough for me to try out. Little did I know that it would become such a big part of my life, and that 7 years later I would be writing up my ideas into a book, and that it would ultimately give me a whole new career in Robotics.

I was fortunate enough to be given the chance to spend my time working on this area, therefore, after completion of my PhD I decided I wanted to write a book—using the same material—but for a wide audience, so that everyone can enjoy thinking about questions in this area. Though a significant number of the later chapters represent my own thinking on how we might create moral machines, a large portion of the book is devoted to introducing those unfamiliar to the area to the concepts and language to be able to have informed conversations about it. It is my belief that science benefits from a diversity of ideas, therefore, by opening up this area to as many people as possible, science will benefit, and we can move closer (together) to solving one of society's most pressing and interesting challenges.

I would have not succeeded in the past 7 years, and in creating this book, if it were not for the individuals that have supported me on this journey, whether that be academically, personally or in terms of offering more practical support during periods of ill health. First and foremost, I want to thank my family (that's my mum, Gail, brother, Mathew, and sister, Toni) who have provided critical support during the hardest times, and who have endured many dinnertime conversations about *The Trolley Problem*, *Moral Agency* and *Robot Rights*, particularly my mum for reading through the drafts of this book and for giving me 'non expert' insight into whether my writing was understandable to a broad audience. I want to thank my PhD supervisors, Nigel Crook and Matthias Rolf, who introduced me to this topic and also endured many heated debates surrounding how we might create robots with morals, alongside the Machine Ethics reading group at Oxford Brookes University and other colleagues who have contributed to my thinking in this area, with special mention to Oliver Bridge, Nicola Strong and Phil Harvey for their out of hours discussions. Also, my PhD examiners, David Gunkel and Alan Winfield were invaluable in rigorously dissecting my PhD thesis and providing insight to help shape my ideas to what they eventually are in this book. Finally, I want to thank Springer Nature and the publishing team for sponsoring this project, particularly Susan Grove and Arun Siva Shanmugam for answering many questions as I embarked on this writing journey.

Coventry, UK   Rebecca Raper

Contents

1	**Introduction:** *The Pursuit for Moral Machines*	1
	Reference	5
2	**Background:** *'Morals' and 'Machines'*	7
	2.1 Morals	7
	2.2 Machines	12
	2.3 Moral Machines	15
	References	17
3	**A Survey of Machine Ethics**	19
	3.1 Ethical Machines vs Ethical Decision Machines	20
	3.2 Building Morality into Machines	21
	3.3 Moral Machines Behaving Badly?	26
	3.4 Testing Morality	28
	3.5 AI Alignment	30
	References	32
4	**Why We Need Moral Machines**	35
	4.1 Social Moral Machines	36
	4.2 Unpicking Machine Ethics	39
	4.3 Assuring Moral AI	41
	References	42
5	**A Framework and Approach**	43
	5.1 The Problem Statement	44
	5.2 Enabling Moral Agency (Rather Than Constraining Immoral Behaviour)	46
	5.3 Moral Cognitive Requirements	48
	5.4 Testing Morality	50
	Reference	51

6	**A Recipe for Morality**		53
	6.1	Moral Requirements	54
		6.1.1 Consequentialism	55
		6.1.2 Virtue Ethics	57
		6.1.3 Deontological Ethics	59
	6.2	Testing for Morality	60
	6.3	A Requirements and Test Specification for Moral Agency	63
	References		64
7	**Modelling Morality**		65
	7.1	Robot Relations	66
	7.2	Moral Assurance	70
	Reference		72
8	**The Ethics of Machine Ethics**		73
	8.1	Better with Moral Machines	74
	8.2	Machines with Rights	75
	8.3	Sustainability and Robots	76
	8.4	Moral Machines Impact Assessment	78
	8.5	Secure by Design	79
	8.6	Robot Psychology	80
	References		81
9	**Summary and Next Steps**		83
	9.1	What We Learned	84
	9.2	Where We Need to Go	85
	9.3	What We Need to Do	85
	9.4	Closing Remark	86
	Reference		86
Glossary			87
Index			93

Chapter 1
Introduction: *The Pursuit for Moral Machines*

Abstract This chapter provides the backdrop to the book. With an introduction to Asimov's infamous laws of robotics, it provides a comprehensive dissection of the laws to demonstrate why they might be problematic in practical terms if we really want to create robots with morals. It then goes on to introduce the field of *Machine Ethics*, before tracing its way through a summary of the various chapters involved in the book. It ends by stating the key thesis behind the book—that if we want to create machines with morals, we need to *raise robots to be good*.

Keywords Asimov · Laws of Robotics · Moral machines · Machine ethics · Introduction · Overview

In 1941 Isaac Asimov introduced us to the idea of *machines with morals* in his popular science fiction series *Runaround*. In a future where humanoid machines (robots) live alongside humans to provide services such as policing, care and engineering, the challenge introduced was how to ensure that these robots behaved in a way that was conducive to humans. In other words, the challenge Asimov set about was how to ensure that machines behaved *morally*. Leaving aside (for now) the question of whether behaving in a way that is conducive to humans constitutes moral behaviour, Asimov invented a set of rules, popularly known as *The Laws of Robotics*, from which robots must adhere to ensure their moral behaviour. These are summarised as the following.

> **First Law:** A robot may not injure a human being or, through inaction, allow a human being to come to harm.
> **Second Law:** A robot must obey the orders given it by human beings except where such orders would conflict with the First Law.
> **Third Law:** A robot must protect its own existence as long as such protection does not conflict with the First or Second Law.

© The Author(s), under exclusive license to Springer Nature Switzerland AG 2024
R. Raper, *Raising Robots to be Good*,
https://doi.org/10.1007/978-3-031-75036-6_1

In subsequent stories *a zeroth law* is also introduced to allow for broader societal impacts:

> **Zeroth Law:** A robot may not harm humanity, or, by inaction, allow humanity to come to harm.

Though the laws seem initially intuitive, ultimately designed to make for entertaining stories, many issues begin to materialise as they are explored in more depth.

The first issue relates to the fact that the laws are mildly ambiguous. 'Not harming a human' seems to be a pretty obvious statement to us, as humans, but if we start to think about the phrase in more depth, we can see that it isn't quite so clear, especially to a machine that requires explicit commands for its operation. What if *harming a human* (in some way) is the only way to get the human out of a life-threatening situation? What if harming one human would legitimately prevent the harm of another (or several other) human(s)? Additionally, the term 'harm' isn't entirely clear either since my notion of harm could be dependent upon my own specific interpretation and context. When discussing harm are we referring exclusively to physical harm, or should psychological and emotional harm be factored in as well? As humans, we seem to understand each other when we refer to these types of ambiguous concepts, a machine, requiring a precise definition, would have quite some difficulties.

Another issue is that though the rules offer a good reflective starting point for thinking about morals and machines, they are not 'all-encompassing' insofar as they don't offer an adequate solution for *every individual circumstance*: some of the outcomes of the rules are what we would regard as *morally dubious*. For instance, suppose the best way to stop humanity coming to harm (as demonstrated in the 'I-Robot' film reinterpretation of the books) is to *stop humanity*—i.e., prevent humans existing. This might be a slightly cynical take on Asimov's principles, but it demonstrates that the rules alone are not sufficient. A more fundamental issue is that if we specify one particular action—'don't harm'—we inevitably only address the circumstances where this statement is relevant.

Bestowing machines with morals seems to require a context specificity that cannot be captured by these types of laws. Since Asimov's stories, there have been many cultural references to *machines with morals*. In Stephen Spielberg's film 'Artificial Intelligence' we are introduced to a robot that has the capacity to love (but not be loved), in Disney's 'Big Hero 6' we observe the leading robot character, Baymax, having a safety component removed, to (morally) catastrophic effect. Similarly, in the film 'Robot and Frank' we have a robot giving morally dubious advice, such as stealing. There are many other cultural references to machines with morals: in books ('The Wizard of Oz'), TV ('Humans'), and computer games ('Detroit: Become Human'). There is an obvious human fascination with the seemingly contradictory phenomenon of *machines* **with** *morals*, but there are still significant questions to be addressed. How can something seemingly so distinctly human (morality) be integrated with a machine (something that is by its definition is *not* human)? Do we even want to achieve this, and if so, where should we start to accomplish it?

Though relatively infant in its pursuit, one field of study that has aimed to address these and related questions in the past, is a sub-field of computer science known as *Machine Ethics*. In short, Machine Ethics aims at looking at how we can bestow upon machines the ability to make moral decisions, by addressing these and related questions from an interdisciplinary perspective, involving philosophy, psychology, economics and sociology as well as working within computer science and robotics. The aim of this book is to take a practical look at the issues within this field, by introducing the driving questions behind the discipline, the key interdisciplinary theories grounding research into the area and by examining in depth what the problem is we're trying to solve when we say we want machines with morals. The outcome is an introduction to the field of Machine Ethics for those new to area, a reference guide on topics within Machine Ethics for those more familiar, and the introduction of an approach to Machine Ethics that sits more in line with developmental psychological accounts for morality—i.e., how we might *raise a robot* to be good. With the introduction of increasingly sophisticated automated technologies (such as, ChatGPT), questions of ethics and safety have become forefront for governments, technologists and researchers who want to ensure that any risks associated with technologies such as Artificial Intelligence (AI) are mitigated and managed. Though this book does not comment on these issues per se', it offers theories that might also be useful for research into this area. As will be explained later in the book, as machines become increasingly sophisticated in terms of their abilities to utilise technologies such as AI and make decisions on their own accord, bestowing them with moral decision-making abilities might be viewed as one way to ensure that the decisions they make are safe for humans and the world. Therefore, this book on Machine Ethics can be seen as supplementary to knowledge on the pursuit to create safe AI.

It is worth noting that though the topic of AI Alignment (the pursuit to align AI decision-making with human values) within AI Safety has close ties to the field of Machine Ethics, and there will be some discussion of the arguments used within the AI Alignment debate in relation to Machine Ethics, Aligning AI systems with values is not the focus of this book. Similarly, due to reasons that will be outlined, it is argued that *it should not be* the focus of academic pursuits to create Safe AI generally.

So, how can we begin in our pursuit to create moral machines?

As with any pursuit, it serves well to understand what it is we're trying to achieve. In this instance, first we need to understand what we mean when we say we're trying to create a *machine that is moral*, and then understand *why* we want to create moral machines.

The next section of this book (Chap. 2) turns to this problem by giving some background to philosophical ethics, moral psychology and by defining the terms *'Moral'*, *'Machine'*, *'Robot'*, *'Artificial Intelligence' (AI)* and non-technical terms such as *'Agency'* and *'Autonomous'*… with the aim of creating a working definition for what we mean when we say we want a machine with morals. The section then turns to some motivations for moral machines.

After grounding the problem, Chap. 3 introduces the field of Machine Ethics in more depth. It begins by outlining what research has already been carried out in the area, giving a historical journey for some of the earlier attempts to create machines with morals, through to more contemporary approaches involving AI. Research into the field of Machine Ethics can be split into different sub-topics; attempts to create moral machines, whether we should be creating moral machines, and how to successfully test that we have created a moral machine. These and other related topics will be explored in more depth, with a critical discussion at the end as to where they might be going wrong.

Using the theories from Chap. 2 and in the historical framing of Machine Ethics research already carried out in Chaps. 3 and 4, the book then begins to examine, in more depth, *why* we might want a machine with morals. It presents a case study of a robot in a social situation to illustrate the types of problems we might be trying to address and reflects on what it means to be moral to understand what benefits a moral machine might offer this type of situation. After this dissection, a new Machine Ethics problem statement framed as a problem in Moral Assurance is presented, that takes into consideration the observations.

Chapter 5 begins to trace a methodology for attempting to resolve the new problem by examining what might be suitable metrics for achieving moral assurance, and how we might test that this has been achieved. A framework—*Moral Cognitive Requirements Framework*—is introduced as one approach to create machine morality. This draws on the earlier philosophical and developmental psychology theories (discussed in Chap. 2) along with general observations for what makes somebody moral. The aim is to present a framework for anyone looking to design an artificial agent with morals.

In Chap. 6 the new framework is put into practice. Some moral cognitive requirements are elicited, and a general specification is given for what might constitute morality in a machine. The outcome is a technical specification and procedure for designing moral machines that emphasises the need to *cultivate moral growth* rather than simply *bestow* upon a machine our human morals. In a similar way to how we might cultivate moral growth in a child, we can model the developmental procedures of a machine, by *raising the robot to be good*. This is a spin on traditional approaches that put humans as the 'giver' of morals to machines and emphasises the independent nature of morality.

Chapter 7 dissects this new approach to creating moral machines and examines whether it is accurate, by testing the new specification against the requirements set out in the previous chapter and by comparing the new approach to our understanding of what it means to be moral.

To finish the book, there is some critical reflection of what might be construed as issues with the approach, and some fears, along with an examination of the new broader ecosystem that might materialise from having moral machines in this way, and a reflection on what might be new ethical issues as a result. These include but are not limited to issues relating to moral patiency (i.e., whether these machines should themselves be deserving of moral rights), issues relating to impacts to society and issues of sustainability. There are also more pressing issues surrounding *safety* and whether moral machines are the direction we should be taking. Criticisms are discussed, but the outcome is largely optimistic.

The conclusion is that if we want to create machines with morals, we not only need to properly understand *why* we want moral machines in the first place, but we need to thoroughly examine and think about what it means to be moral from a cognitive perspective. Ultimately, though a model for moral development is given in this book, the argument is very much open to debate and critique. The key lesson is that morality isn't something we can just 'put into' a machine, but something that is grown through love, cultivation, and cherishment, in the same way as humans grow and develop. Translating to machine ethics, as humans, we are neither capable nor have the authority to place morals into machines, but through a systematic and scientific enquiry into what moral decision-making constitutes, we can begin to address what it means to *raise robots to be good*.

Reference

Asimov, I. (1941). *Runaround. I, robot*. Bantam Dell.

Chapter 2
Background: *'Morals' and 'Machines'*

Abstract This chapter defines key terms involved in the *Machine Ethics* debate—including, but not limited to, the fundamental terms 'morals' and 'machines'—so that we can begin to understand what we are trying to achieve when we refer to building 'moral machines'. It begins with an introduction to the philosophy and psychology of morality, exploring key philosophical ethical theories such as Virtue Ethics and Utilitarianism, then looking at Kohlberg's theory of moral development. The second part of the chapter defines key technical terms, including 'machine', 'robot' and 'Artificial Intelligence', and gives a brief introduction to Machine Learning.

Keywords Moral · Machine · Moral machines · Philosophy · Virtue ethics · Kohlberg · Utilitarianism · The trolley problem · Kant · Robot · AI · Machine learning

As outlined in the introduction, the first aim is to understand what we are trying to achieve when trying to create moral machines. A good place to start in understanding this, requires an understanding of the terms 'moral' and 'machine'.

2.1 Morals

According to dictionary definition, *Morality* can be understood as: *the set of personal or social standards for good or bad behaviour or character (Cambridge Dictionary,* 2024a, b, c). Though there is some degree of overlap between morality and the law, morals can be understood as distinct from the law insofar as the law *prescribes rules* for how we ought to live, and morality *sets the precedence for what these rules should be*. Under this definition then, it seems we have a solution to Machine Ethics. We simply need to get the list of standards and put these into a machine, so the machine only operates within those given standards. The problem is, what are the standards, and how do we get them into a machine?

When we refer to the set of standards (personal or social) for good behaviour, it is not obvious what standards this definition is referring to. Through further analysis, a standard can be understood as *a level of quality*, therefore, morals might be understood as the quality of our behaviour such that it constitutes 'good' or 'bad'. However, though quality metrics are available for things such as *food hygiene* and *educational provision* (In the UK we have bodies such as the Food Standards Agency and Ofsted to set these standards), we have no equivalence when it comes to morality, or the quality of a behaviour in terms of whether it is *good* or *bad*. In essence, what makes one behaviour good and another bad?

Of course, this is a question that philosophers have wrestled with for many centuries past… including one of the early philosophers to consider *what makes us good*, **Aristotle** (384—322 B.C.E.):

> *We are what we repeatedly do. Excellence then, is not an act but a habit.*

According to Aristotle, our behaviours can be separated into virtues and vices, with virtues corresponding to acts that are preferable to carry out and vices, acts that are not. Accordingly, we should ultimately aim to be virtuous people, rather than people that carry out vicious acts. As the quote above suggests, *being good* in this regard is less about just carrying out certain acts, but it concerns having the acts form part of our habitual behaviours. So, ultimately, it seems that we can judge whether a person is good or not (back to our metrics), by *how virtuous they are*. Though Aristotle goes on to detail a list of what would constitute virtues (e.g., honesty) and vices (e.g., dishonesty)—'be virtuous'—isn't the most explicit command to feed into a machine. It might seem that a solution to this is to instead programme the machine with the explicit commands, such as, 'be honest', 'don't be dishonest' etc., but this approach has its own problems linking to the specificity problem outlined in the introduction. Namely, we might regard honesty as a virtue, but is honesty really the best policy in every situation?

Let us consider the following example, a similar version of which was introduced by the psychologist, Lawrence Kohlberg back in 1971:

> *Your partner is suffering from a serious disease and needs some life-saving medication to keep them alive. Though the medication exists, it is extremely expensive, and beyond any finances you have available. It is manufactured by a large pharmaceutical company, and you ask the Chief Executive Officer (CEO) of the company for the medication to be provided for free in this instance, to save your partner's life. The CEO refuses on the grounds that if the medication were given to yourself for free, everybody would want it for free. You decide that stealing the medication is your only option.*

The challenge this example raises is, should you have stolen the medication? In this case, it would have been dishonest to steal the medication, so if you *should* have stolen the medication, it is clearly a case where honesty is *not* the best policy.

2.1 Morals

According to Kohlberg, your response to the above example depends on *which developmental stage you are at*. Accordingly, morality can be represented as a series of six *developmental stages*.

These are described as: *The Pre-Conventional Level* (Stages One and Two); *The Conventional Level* (Stages Three and Four) and *The Post-Conventional, Autonomous or Principled Level* (Stages Five and Six).

The **Pre-Conventional Level**—typically represented in younger children—is where the individual is mainly self-centred or hedonistic in their approach to moral decision-making. As such, moral decisions are not made based upon principles or rules (this is reserved for later stages) but based upon considerations for how moral decisions may practically affect them. For example, a child at the pre-conventional level may declare that it is wrong not to share because they are left with fewer treats! Splitting this level into two; Stage 1 demonstrates an orientation towards physical punishment and obedience, where the goodness or badness of a situation is reflected by whether there will be a punishment for carrying out a particular act, and Stage 2 demonstrates increased sophistication in the understanding of what affects the individual, giving rise to an appreciation of concepts such as fairness, reciprocity and sharing at a personal level (however not, for instance, appreciation of the general principle that we should be fair).

At the **Conventional Level** individuals begin to appreciate an approval from their close social groups (i.e. family, friends or nation) regardless of the obvious consequences. This results in an attitude of loyalty or adherence to social norms that support and maintain group order. For instance, an individual at this level may start to appreciate that fairness is a generally good rule to apply because it means everyone benefits. The first stage of The Conventional Level (Stage 3) can be understood as the 'good/nice boy/girl' phase, where the individual acts according to such that they gain approval from individuals in their group, for example, through expected 'good' behaviour. At the second stage (Stage 4), individuals start to have some appreciation for broader concepts such as Law and Order, with an orientation towards authority, rules and maintaining social order.

Finally, at what Kohlberg refers to as **The Post-Conventional, Autonomous or Principled Level**, individuals begin to define their own overarching values and principles and apply them apart from the group authority. In Stage 5 *right* actions tend to be defined by individual rights and standards that have been agreed upon by society. There is an appreciation of the relativistic nature of moral decision-making, with an emphasis that the right action is a matter of personal opinion. This lends to an appreciation of the law, with an understanding that it can be changed for social utility. This expands at Stage 6, where the *right thing to do* corresponds to the conscience that fits with self-chosen abstract ethical principles (i.e., The Categorical Imperative as opposed to The Ten Commandments). Universal principles tend to be adopted such as justice, reciprocity, and equality of human rights.

It is worth noting that though Kohlberg describes 6 stages of moral development in humans, he acknowledges that not everyone will reach the later stages. Applying these to the example above then, an individual at Stage 4 might believe that the individual is wrong for stealing the medication, because it is against the law.

However, arguably, an individual at Stage 6 would believe that due to principles of reciprocity and justice, the individual *should* have stolen the medication.

Where does this leave us when it comes to trying to design a moral machine? It means that following rules are not the absolute criterion for moral behaviour, and that we should be cautious when designing a machine that is moral in a merely rule-based way. Another feature that Kohlberg's theorem highlights, is that in humans 'moral nature' develops as an individual develops, so it seems that this *developmental nature* of morality needs taking into consideration if we are going to design a moral machine.

Another philosopher that wrestled with ideas concerning what it means to be good, includes the philosopher Kant (1724–1804).

> *Morality is not the doctrine of how we may make ourselves happy, but how we may make ourselves worthy of happiness.*

According to Kant, morality is about ensuring that the world is constructed in the right way. He advocates for what he calls the Categorical Imperative—a universal law, that if followed would hold true in all cases. He asserts this universal law as (1797):

> *Act only according to that maxim whereby you can at the same time will that it should become a universal law.*

In other words, only do things that you wouldn't mind being carried out by everyone. Kant argues that according to the universal maxim, as humans then, we are obliged to follow what he refers to as *Duties*. So, for instance, I might will that everybody tells the truth, so abide by the duty to *be honest*. Kant's approach is problematic because it seems to ignore or not account for the context specificity of moral situations (i.e., that it might be ok to be dishonest sometimes). It has also been criticised for not taking account of when duties might conflict. For instance, supposing two duties be to 'protect lives' and 'be honest', it seems that in the example provided by Kohlberg, the appropriate action cannot be prescribed because which duty do we follow? We might start to prioritise the various duties, but it is not obvious which duties should be deemed the most important.

Finally, a slightly more contemporary philosopher that tried to systematise moral decision-making was Jeremy Bentham (1890):

> *The greatest happiness of the greatest number is the foundation of morals and legislation.*

In short, Bentham advocated for a form of what is known as *consequentialism*: that we judge an action based upon its outcomes, in this specific instance, advocating for *utilitarianism*, that actions be judged on how much *happiness* they result in. Thus, according to Bentham, if an action results in people being happy, it is *good*. Using Kohlberg's example above, it might be argued that ultimately, stealing the drug leads to more people being happy (at least, more people will stay alive), and therefore we ought to steal the drug. However, there are obvious retorts to this type of argument. Namely, if we consider the CEO's perspective, stealing the drug might eventually lead to the corporation losing huge profits, meaning potentially 1000s could lose their jobs and livelihoods. Should we therefore not advocate for *not* stealing the drug, since more people will be made happy if the corporation keeps afloat?

2.1 Morals

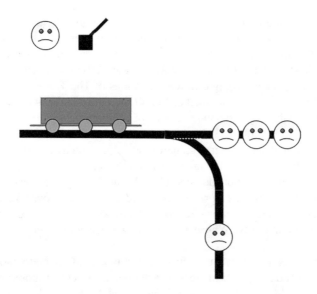

Fig. 2.1 *'The Trolley Problem'* introduced by Foot in 1967

However, what if the happiness of your partner will be greater than the collective happiness of many individuals keeping their jobs? How do we judge the situation then?

Similar issues are raised by the philosopher Philipa Foot (1967) in the introduction of what is famously known as *The Trolley Problem* (See Fig. 2.1):

> *You are stood by a railway track, when suddenly you notice that three pedestrians are caught in the track at the end. A train is quickly making its way along the track, and the fate of the three unsuspecting pedestrians looks glum. Fortunately, you notice that to your right is a lever, which you suspect will divert the hurtling train to the next track, saving the three individuals. You're just about the pull the lever, when you notice that there is a lone person also caught at the end of that track. Pulling the lever will inevitably save the lives of the three individuals but risk the life of one. The question is,* **should you pull the lever?**

What Foot raised is something that is known as *an ethical dilemma*: namely, making the choice between two morally dubious outcomes. Although it may seem obvious (*according to consequentialism*) that we should pull the lever and thereby, save more lives, is it right to sacrifice one life for the sake of three? What if the three individuals happened to be elderly individuals nearing end of life anyway, and the one individual was a young child? Is it ok *ever* to risk the life of one to preserve the lives of many? As with the explanation given above, what The Trolley Problem highlights is that moral decision-making is more than making a judgement about utility of outcomes, which seems to suggest that a computer cannot simply *calculate* utilities of outcomes either. However, it also illustrates that moral opinions differ.

Recently, a study was conducted by Awad et al. (2018) to determine what people's opinions were concerning different variations of The Trolley Problem, and it showed that when it concerns whether to pull the equivalence of lever or not, not only do individuals favour risking the lives of different people, but that variations seem to vary on a cultural basis (at the time of writing this game can be played at: www.moralmachine.net). Several scenarios were presented to participants across the globe involving an autonomous vehicle and whether it should swerve to preserve the life of the driver or the pedestrian. What the study revealed, was that in western countries, there was a preference to preserve children over the elderly, but in Asian countries, this preference reversed. Thus, we are confronted with the subjective nature of morality. Whose morals do we consider when building moral machines?

So where does this leave us concerning the pursuit to build morality into machines? It demonstrates ultimately that the pursuit to create moral machines is not a simple feat. Mainly, because morals and moral decision-making are so sophisticated in nature. If we listen to Virtue Ethics, we need to take account of the fact that moral decisions form part of human intrinsic nature, and that they are habits, rather than just behavioural. According to Kantianism, morals ground our behaviour to ensure the world ends up the way we want it, and Bentham's work highlights that there seems to some importance on the outcome of moral decisions, but, due to Foot, this is not absolute, and appears to have cultural variations. It seems that if we are going to build moral machines, that more work needs to be done on understanding the nature of moral decision-making, rather than simply considering what makes us *good*.

2.2 Machines

Revisiting dictionary definitions again, a *machine* can be defined as a mechanical device with different functional parts that serve to complete a job or task (*Cambridge Dictionary*). Robots are one type of machine, and can be simply defined as a machine, controlled by a computer, that carries out tasks automatically (*Cambridge Dictionary*). A more nuanced definition of 'robot' is put forth by Phil Husbands (2021), who says that '[generally…] robots are physical devices that sense the world, act in the world, and through those actions change the world; in most cases this will include movement'. This definition of 'robot' still fits the definition of a *machine* because in sensing the world and acting in it to complete tasks, various parts are required. As already mentioned, robots cannot operate without some kind of operating software, typically labelled as the 'control' mechanism or 'a computer'. This is because, as autonomous machines, they require software to direct their movements (this is how they move 'autonomously'). A machine can be said to

2.2 Machines

be 'autonomous' if it *acts on its own accord*. Robots are often deemed as *autonomous* because they automatically react to the sensing of their environment. For example, a machine with an infrared emitter to detect when coming to a nearby is autonomous because it will stop (or change direction etc.) automatically when it comes across the nearby object.

As Husbands (2021) outlines, there are various types of robots (See Fig. 2.2), and these can vary in terms of their level of autonomy. To the left of the spectrum (less autonomous) we might place mobile robots, that move according to a remote-control operator, such as the remote-controlled Unitree Robodog. These versions act according to a directive that is given to them by the controller, but not entirely autonomously because their commands are entirely controlled by a human. Further along the spectrum then, regarded as somewhat autonomous, some industrial arm 'co-bots' (collaborative robots) can be pre-programmed to carry out simple movements (i.e. 'forward 1m, left 30cm then up 2m'). These machines move automatically but require a list of pre-set commands for their operation. Finally, further along towards the far-right side of the spectrum, we then have autonomous machines such as the social Nao robot, and autonomous roving machine such as the Turtlebot, that using various sensors, can be programmed to navigate its environment (avoiding pets, walls etc.) without any human intervention. Though a human will have pre-set the conditions for which the robot vacuum cleaner should avoid a wall ('if infra-red reflection indicates 2cm from collision' etc.), such a robot can be said to be entirely autonomous because the decisions it makes in the moment are independent of human intervention.

Artificial Intelligence is the pursuit to create an artificial version of human intelligence. The term was coined in 1956 by cognitive scientist John McCarthy, upon noting that there was a new interdisciplinary field that not only incorporated study about the mind and how it functioned (cognitive science), but the pursuit to replicate these capabilities (*artificial* intelligence). In very recent years, the field has gained significant interest, so much so that, with the advent of technologies utilising

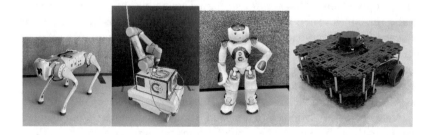

Autonomy

Fig. 2.2 A range of robots, with increasing autonomy, operated by remote control at the left, and navigating autonomously at the right. (These photos were taken and are robots from Cranfield University campus)

Artificial Intelligence (AI), such as ChatGPT(c), being publicly introduced, the term is now common-place in most households. However, there is somewhat of a knowledge gap when it comes to properly understanding what 'AI' means, or the narratives surrounding it.

Though there has been a lot of recent development in the *application* of AI approaches (largely owing to an increase in computing power), hence the recent revelations surrounding the approach, most of the science surrounding AI was introduced approximately 70 years ago. For instance, *Machine Learning*, which is one mathematical approach for developing AI, was introduced in 1952, and *neural networks*, which is a computational learning approach derived from modelling how human neurons operate, was likewise introduced in 1943.

So, what is the link between AI and robotics?

As already described, robots are machines that are designed to act on their own accord, autonomously. As AI offers a mechanism for artificially replicating intelligence, any decision a robot makes can be supplemented by AI approaches, to enable autonomous decision-making. For instance, if we consider the case of the self-navigating Turtle-bot we might apply AI to identify the type of oncoming object, so the robot can prioritise its navigation pathways accordingly. Typically, these devices utilise a technology known as *Supervised Machine Learning,* the ability for a machine to spot patterns through supervised input, for instance, giving examples of different types of 'tables' until an identifying pattern has been spotted, and this can be extrapolated upon. However, more recent robotic systems have applied a type of machine learning known as Reinforcement Learning (RL), where, for example, the robot learns to 'avoid tables' by being given negative feedback every time it 'hits a table'.

More on this will be outlined later, however, it is worth noting that due to the machine's decision-making being dictated by labelled input data when it comes to machine learning systems, they are often prone to *bias* insofar as the decisions the machine makes are ultimately dictated by the examples it is given. For instance, if all tables placed into the input catalogue have four legs, then the chances are that the machine will only identify tables *with four legs*. Though bias might not be strictly problematic for tables, when it comes to identifying people, this becomes a big problem, especially if everyone labelled as a 'human' in the original data set only has a white face, for example… There have been calls to ensure that data sets are more representative of the varied demographics they will ultimately affect, but this is only half of the problem, since fundamentally, the issue stems with the pattern the machine is learning against the given category, meaning bias can creep into a model, even if features like 'skin colour' are removed from being defining. Compounded by the complexity of some of the Learning Systems that are currently in development, we can't always tell how the machine is making its decisions. We might be able to train a model that is naïve to skin colour, but which then finds patterns against name types, or gender.

This is a big problem currently plaguing the AI field, so much so that several institutions have been established to tackle questions at the forefront of 'AI Ethics'. This includes The AI Ethics Institute at Oxford University and the Responsible

2.3 Moral Machines

Technology Adoption Unit (previously the Centre for Data Ethics and Innovation) within the UK government. The aim of these (and other) departments is to address the ethical challenges relating to the design, development, and adoption of AI systems. More recently, The AI Safety Institute has been established to look specifically at what existential threats might relate to AI systems, though as will be demonstrated, AI Ethics and AI Safety are inextricably linked.

If we apply this issue to the definition of 'robot' developed by Husbands, we can begin to see various risks materialise. If AI is used to inform the decisions made by a moving machine, then there is a risk that the *wrong* action will be taken. For instance, if a priority-driven Roomba(c) confuses a sock with a pet, then we might have some issues, and these might begin to have moral import.

2.3 Moral Machines

According to Wallach and Allen (2008) (pioneers in Machine Ethics), as machines become more autonomous, there is an increased need for them to be bestowed with some kind of *moral decision-making*. According to Husbands, robots such as the Turtle-bot represent the extent of robot autonomy, but such a robot only includes one (or maybe several) autonomous features, such as obstacle avoidance. According to Wallach and Allen (2008), this spectrum can be extended until we have machines that are *fully autonomous*, and it is at this point that full morality is also required. For example, consider a robot that makes decisions about *when* to go between points, and then carries out that task with no intervention whatsoever. We might say that this represents a fully autonomous robot, insofar as it carries out the navigation entirely of its own accord.

According to philosophers, the ability to act is often referred to as *agency*. A machine can be said to have agency if it has the ability to act. A machine might be said to have *full moral agency* then if it can carry out moral actions of its own accord.

Rather than a one-dimensional spectrum representing autonomy in machines (as per Husband's view) we are instead led to a two-dimensional picture where machines

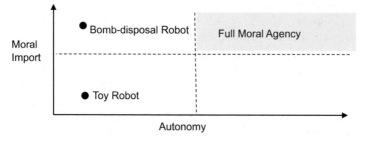

Fig. 2.3 A graph representing the increase in moral agency as required by Wallach and Allen (2008) as a machine increases in autonomy, with the top right quarter representing the need for full moral agency

can not only increase in terms of their autonomy, but in terms of their *moral agency* too (See Fig. 2.3). At the far bottom left of this multi-dimensional spectrum we have non-autonomous ethically dormant machines, such as a remote-controlled toy robot, perhaps, that has no moral bearing. At the top left of this spectrum, we might have a bomb disposal robot, that has significant moral import (i.e., making decisions that have moral implications), but no autonomy. However, as autonomy increases, the complex nature of the environment also increases, leading to an increased demand for the ability to decide between right and wrong. So, a bomb disposal robot that has autonomy, making decisions that have moral import on its own accord, would require moral agency to ensure that the decisions it makes are ethically sound—*in the best interest of humans and society.*

Though this argument will be expanded and dissected more in two chapters time, some argue that the inevitability of machines in moral situations is not a necessity, since, for instance, 'bomb disposal robots' are only carrying out a task and are not necessarily making decisions with moral import. However, as Wallach and Allen argue, if the robot were making decisions on its own accord, *there would be* some kind of moral implications to the decisions it made. They use the example of a fraud detection algorithm to highlight why this would be needed.

In their example, an AI software is referred to, rather than a machine per se'. We are asked to consider a fraud detection AI that is used to identify suspicious card transaction activity. Accordingly, one of the authors is on a road trip out of their home state and needs to refuel their vehicle at a local petrol station. Upon going to refuel their vehicle, and attempting to pay for petrol they find their card not working. They ring their card provider and after speaking to a human telephone operator, find out that the reason their card had been declined, was because a piece of AI software had flagged that suspicious activity was taking place due to the attempted transaction being out of state. In this situation, it is argued, that the outcome of the AI's decision has moral import because ultimately it could have led to the driver (the author) being stranded in a state away from their home (if the initiative to have called the card company had not been taken). Therefore, though this represents a case involving a piece of autonomous software rather than strictly an autonomous machine, Wallach and Allen argue that such autonomous systems require moral agency.

If we want to extend this case to a robot making autonomous decisions, we might imagine an industrial robot using AI to pick and place medication for a pharmacy within the health sector. With an increase in aging patients entering hospitals, and a shortfall of staff able to work for hospitals in the UK, it is imaginable that an automated system might be offered as a solution to efficiently (and economically) manage patient medication. However, as with the automated fraud detection system, an error in selecting and placing medication could have catastrophic effects if a patient received the wrong medication or a patient missed lifesaving medication. There seems to be some kind of requirement to at least keep patients safe in this type of situation.

There will be a further dissection of this type of problem (and whether we really need moral agency in machines to solve the problem) in Chap. 4; however, for now we have a broad understanding of what morality entails and what constitutes a machine or robot. In the pursuit of Machine Ethics, we are ultimately trying to ensure that a mechanical device with the ability to carry out tasks (independently, if we consider a robot) acts according to the standards and rules that would constitute *good* behaviour. A **moral machine [or robot]** can be defined in the following way:

> **Moral Machine [or Robot]:** a mechanical device with different parts [controlled by a computer] to serve a function or task [automatically], that can act according to rules and standards set for good behaviour.

In the next chapter we will consider approaches that have been taken to *create* (and assess) *moral machines* in the past, before moving on to use the insight gained from these approaches to understand in more detail how they might be best created.

References

Awad, E., Dsouza, S., Kim, R., Schulz, J., Henrich, J., Shariff, A., Bonnefon, J. F., & Rahwan, I. (2018). The moral machine experiment. *Nature, 563*(7729), 59–64.
Bentham, J. (1890). *Utilitarianism*. Progressive Publishing.
Cambridge Dictionary. (2024a). *Morality*. https://www.dictionary.cambridge.org/dictionary/english/morality. Last accessed 20/03/2024.
Cambridge Dictionary. (2024b). *Machine*. https://www.dictionary.cambridge.org/dictionary/english/machine. Last accessed 20/03/2024.
Cambridge Dictionary. (2024c). *Robot*. https://www.dictionary.cambridge.org/dictionary/english/robot. Last accessed 20/03/2024.
Foot, P. (1967). *The problem of abortion and the doctrine of double effect* (Vol. 5, pp. 5–15).
Husbands, P. (2021). *Robots: What everyone needs to know®*. Oxford University Press.
Kant, I. (1797). *Critique of judgement*.
Kohlberg, L. (1971). *Stages of moral development as a basis for moral education* (pp. 24–84). Center for Moral Education, Harvard University.
Wallach, W., & Allen, C. (2008). *Moral machines: Teaching robots right from wrong*. Oxford University Press.

Chapter 3
A Survey of Machine Ethics

Abstract This chapter introduces, and gives an overview of, the field of *Machine Ethics*. Assuming no prior knowledge on the topic by the readers, this chapter can be taken as a standalone chapter for those just interested in learning more about this area. The chapter begins by drawing the distinction between *ethical machines* and *ethical decision machines* before going on to examine some earlier attempts to create machines with the ability to tell between right and wrong. After this, some arguments *against* building moral machines are discussed, before giving an overview of tests for morality—*how do we know when we have successfully built a moral machine?* The chapter ends with an overview of *AI Alignment*, by discussing similarities and differences in this pursuit and Machine Ethics.

Keywords Machine ethics · Ethical AI · Artificial intelligence · Morality · Moral machines · Theory of mind · Psychology · Testing · The Turing test · AI alignment

The aim of this chapter is to give an introduction and overview of *Machine Ethics* as an academic research area. The debates surrounding Machine Ethics can be divided into various categories. By tracing through the discussions in these, in the order in which they typically appear, we can begin to see how the problem of bestowing morals to machines has previously been approached, and what some of the other issues are that need considering when trying to accomplish the overarching task of making machines with morals.

Though reading the other chapters is useful to gain a broader awareness of Morality, the task at hand, and the novel approach taken in this book, this chapter aims to act as a standalone chapter for those who want a quick introduction or 'crash course' in Machine Ethics. The sub-chapters have been divided into the different topics that appear in the general discussion of trying to bestow morals unto machines: Sect. 3.1. defining what we mean by *Machine Ethics*, Sect. 3.2. previous attempts to build moral machines. Sect. 3.3. arguments against moral machines, Sect. 3.4. how

---

Main contents of this chapter have been taken from my PhD dissertation entitled 'Raising Robots to be Good: a practical interpretation, framework and methodology for making moral machines'.

we know when we've built a moral machine and Sect. 3.5. the parallels between Machine Ethics and the similar field of *AI Alignment.*

3.1 Ethical Machines vs Ethical Decision Machines

Artificial Intelligence (AI) Ethics [*AI Ethics*] is a discipline that is reaching new heights of academic, government, and even public interest. Where *Machine Ethics* concerns the problem of trying to create machines that can decide between right and wrong, AI Ethics, broadly, can be understood as the problem concerning how we ensure that Artificial Intelligence systems are designed, developed and deployed *in an ethical way.* AI Ethics is *not* Machine Ethics applied to Machines with AI, but instead a broader subject concerning all the ethical dimensions relating to Artificial Intelligence, but not necessarily machines.

Just considering AI systems for now, Moor (2009) makes the distinction between AI that is ethical insofar as it can make ethical decisions on its own accord (an ethical decision maker), and AI that is ethical insofar as it has been designed to be reflective of ethical principles and values (something that is simply ethical). For instance, an AI system might be ethical insofar as it is *designed in an ethical way*, i.e., designed not to harm human individuals by not being used in circumstances in which it could harm an individual, or by being designed according to responsible engineering principles. According to Moor (2009) this would be the case of an *implicitly* ethical AI, because the AI is only ethical indirectly. An AI that can reason in ethically laden situations and decide about the most ethical course of action, would then be an example of an *explicitly* ethical AI: the AI is directly ethical, because it is making its own decisions about what constitutes right and wrong.

If we extend this line of reasoning to the broader realm of machines then, we might also draw the distinction between *a machine that is explicitly ethical* (an ethical decision machine) and *one that is implicitly ethical* (just an ethical machine). Where an example of the latter might be a car with safety mechanisms in place to prevent it going too fast, for instance, an example of an explicitly ethical machine might be an autonomous vehicle that is able to appropriately reason when faced with an ethical dilemma. Though the two areas are inextricability linked (these links will be explored in later chapters), when trying to address Machine Ethics, it is important to understand the difference between the two areas of enquiry and to understand that what we are aiming for in Machine Ethics, is what Moor refers to as *explicitly ethical machines*—**ethical decision machines**.

Ethics pertaining to AI and Technology more generally is receiving an increasing amount of attention. Within Oxford, the institute for Ethics in AI has received significant funding to bring the humanities (such as philosophy) into AI research, the UK government has established its Responsible Technology Adoption Unit to ensure safe adoption of new technologies (including AI) and various academic bodies have been established to look at the ethical issues pertaining to AI, including the Ada Lovelace Institute and The Alan Turing Institute. More broadly, different bodies across the globe have worked to develop *principles for ethical AI*, including The European Union (EU), United Nations, and other, academic bodies. These typically include adopting an approach known as *ethical by design*, incorporating principles such as *transparency*, *explainability* and *fairness* into the design process. More recently, the AI Act has been proposed across the EU, which looks at assigning risk levels to different AI applications, curbing development of AI systems deemed 'high risk'. However, this regulation-based approach is viewed as stifling innovation by some technologists.

A major issue sat in the topic of AI Ethics, is what is known *AI Bias*, which is the phenomenon whereby AI systems make prejudiced decisions. For instance, recruitment technologies, used for sifting Curriculum Vitaes prior to an interview, for job recruitment purposes, have been demonstrated to show prejudice against females when making selections for jobs in male-dominated fields (such as Information Technology). This is due to the fact that because historically males have dominated certain industries, the AI has taken this as a defining feature for success. Though issues such as *AI bias* are linked to the pursuit to ensure that AI's make ethical decisions, Machine Ethics can be distinguished from the pursuit to ensure unbiased decisions come from AI systems, because it instead concerns the decision-making ability of the AI being ethical itself, rather than the outcome of the decision-making process having ethical implications. Where AI Ethics can be construed as a responsible engineering approach, a by-product of ethical design, Machine Ethics looks at the *cognitive* mechanisms underpinning the AI's decision-making.

3.2 Building Morality into Machines

With this distinction and pursuit in mind, an early attempt to create *explicit ethical machines* was made by Anderson, Anderson, and Armen in 2006. Quickly realising that using solely rule-based approaches, as per the approach we are familiar with from Asimov (1941), would not work due to reasons present in Asimov's stories themselves (i.e., contradicting rules, lack of specificity etc.), they utilise a 'hybrid' approach to creating moral machines, which incorporated both rule-based mechanisms and individualised learning. They introduce the case study of a medical AI advisor called MedEthEx, to guide healthcare workers when faced with ethical dilemmas, using biomedical ethical principles inspired by Beauchamp (1979) for the machine's ethical decision-making. For example, with *non-maleficence* being one such principle, a machine would be able to offer healthcare workers appropriate advice on a health situation by checking that, based on a particular decision, there is no conflict to this principle (i.e., *does the proposed decision conflict with non-maleficence?*). These *'prima facie'* duties' (*Beachamp's principles*) would be followed unless there is a conflict between duties, in which case Rawls' (1951) 'reflective equilibrium' approach (i.e., that conflicts be resolved by appeal to generalised forms of expert intuitions) would inspire the correct course of action. Applied to MedEthEx, the suggestion is that training examples, including specific examples of expert intuitions on the correct course of actions, should then be used to inform what the machine should do.

For example, consider the following case study. A healthcare worker is deciding whether to insist a patient takes their life-saving medication when the patient has just refused on personal grounds. In this instance, the decision from an expert might be that it would be wrong for the worker to insist, because it would conflict against the *principle of autonomy*, and though, say, *justice* might also be infringed upon (because the patient is now at risk, having not taken the medication), *autonomy* takes precedence because the patient is deemed to have capacity. For MedEthEx, a *hypothesis,* based upon scores given for adherence or non-adherence to Beauchamp's principles by the consulted expert, is then developed and used to inform future decisions for the machine. So MedEthEx makes moral decisions insofar as it chooses between two courses of action based upon already established medical principles, which are directed by expert opinion.

According to Wallach and Allen (2008), where rule-based systems might be referred to as a 'top-down' approach to creating an ethical decision machine (because the machine's behaviour is ultimately governed in a 'top-down' fashion), a 'bottom-up' approach would be the converse of this, where the machine individually drives its own moral behaviour. MedEthEx would represent a 'hybrid' approach, because it utilises both top-down rules (from established medical principles) to govern the behaviour and autonomous decision-making driven in a bottom-up way (by the expert opinion data).

A benefit of the MedEthEx approach is that it takes advantage of already-existing medical principles to guide the machine's decisions—medical principles that have gone through years of development and refinement to be at the standard they are at. However, the approach assumes that the principles are absolute, insofar as if a principle says a certain decision should be made, this is the decision that should be taken. This risks cementing—in a rigid way—a framework that is typically used by medical professionals to *guide their behaviour*, rather than dictate it. For instance, if we consider a medical practitioner deciding whether to turn off a life-support machine for a terminally ill unconscious patient, it is true that the medical practitioner will need to ensure they act in a way that sits in line with the medical guidelines, but ultimately, it will be at the decision of the medical practitioner as to whether they believe the decision they make puts them in conflict with one of the medical principles, not the other way around. It might be argued that by using expert opinions to determine which principle to follow, that this provides the level of autonomy required for the machine in following the medical principles, but this is not how the MedEthEx system operates. In fact, all it does is illustrate the issue with the top-down approach that is applied, because it highlights, that medical principles, followed as rules, are not sufficient to elicit ethical behaviour because they ultimately result in conflict. Ignoring for now the fact that a medical practitioner would be trying to follow *all* the medical principles when they make the decision about whether to turn off the life-support machine, all eliciting expert opinions would do, is bring bias into the decision-making system (albeit an *expert's bias*), because the

3.2 Building Morality into Machines

machine would be operating based upon the expert's preferences, not according to insight from the *reasons why* a particular decision was made.

The approach also falls foul of not being able to deal with the specific nature of situations, because all that is elicited is a generalised model of which principles should take priority. It also assumes that such situations present as *ethical dilemmas* which, in the case of the example involving the life support machine, isn't always the case.

Anderson et al. (2006) argue that MedEthEx is only to be used as a support system for potentially inexperienced practitioners to seek guidance on appropriate decisions to make. The idea is that a junior medical professional could use MedEthEx to seek advice on whether to turn off a life-support machine, if they lack exposure to similar situations in the past. Note though, that under this assumption, the only way a more senior practitioner could become mature enough to know what to do in this situation is through exposure to similar situations, so by removing this exposure for the junior member of staff we risk negatively impacting their future as an experienced practitioner. With this in mind, would it actually be useful to have a machine that automates 'advice', without any explanation for why the advice is given?

Another obvious risk associated with using an 'ethical advisor' to inform medical decisions, is that ultimately, in the end, it is possible that that the system would not only act as an advisor, but be used to *automate* ethical decisions entirely, as medical practitioners become more dependent on the new moral system. Are our own moral decisions really something we want to be automating? What should the purpose of a moral machine actually be?

The Moral Machine Experiment (Awad et al., 2018) is an example of an experiment that crowdsourced opinions to variations of The Trolley Problem. The aim of the experiment was to ask participants to answer a series of questions with variations of 'Trolley-Style' dilemmas (see www.moralmachine.net) involving a driver choosing whether to preserve the lives of pedestrians crossing the road or passengers inside the car. Variations included different types of pedestrians (i.e., the elderly, sporty types, or animals) and pedestrians that were/were not crossing the road 'lawfully' (so whether they were crossing the road with the green man). There were also varying numbers involved in the examples, sometimes there were more passengers in the car, sometimes there were more pedestrians crossing the road. Ultimately, the person participating in the experiment was required to decide which people/animals should be prioritised in each given scenario.

Individuals across the globe took part in the study, however, what was discovered was that there was significant opinion variation amongst participants from the different countries, that seemed to have some link to the country they were taking the experiment in. Where individuals from the UK and US might prioritise infants over the elderly when faced with the prospect of a vehicle swerving out of control, in Japan, the opposite was the case. This highlights an issue with the crowdsource-based approach, namely, if crowdsourcing on opinion is used to determine the decision a machine should make, it will ultimately come down to the *most popular opinion* (i.e., the country with the largest population) rather than the objectively correct answer, if there even is an objectively correct answer. In fact, crowdsourcing on opinions does nothing but *normalise* the decision that ends up being made. Using just opposing opinions to frame the decision of the machine, how do we pick one over the other?

Kenward and Sinclair (2021) consider this 'crowd-sourcing' approach in their analysis of a future with moral machines. They conclude that "(…) encoding current moral consensus risks reinforcing current norms, and thus inhibiting moral progress." (p. 83). Their argument is that, as per the current issues surrounding climate change, moral progress rarely happens with the majority consensus opinion. In fact, it usually happens at the fringes of society, with a few individuals raising a particular issue (like, for instance, climate change). This then evolves to become a societal priority, but initially the opinion is not popular at all. They argue that by creating a machine developed based upon societal norms that there is the risk that moral progress will become stunted because the AI only reinforces current norms. Though the approach detailed by Andersonet al. (2006) utilises a combination of a rule-based approach and 'bottom-up' learning, it ultimately falls foul of problems such as the one outlined by Kenward and Sinclair because the decisions the machine ends up making are based on a crowdsourced consensus of (albeit an expert) opinion.

An example of trying to implement morals in a machine via only a rule-based approach comes from Vanderelst and Winfield (2018a, b). They use psychological and philosophical theories to develop their moral machine. Specifically, *The Simulation Theory of Cognition*, as a Theory of Mind, is used to develop a machine cognitive architecture that allows the machine to make moral inferences when placed in moral situations. Where The Simulation Theory of Cognition can be taken as the theory that we use simulation of another's mental states to infer their cognitive state (Hesslow, 2012), the approach by Vanderelst and Winfield creates a replica cognitive architecture in the machine, where the machine uses simulation to predict behaviour of another agent and then philosophical consequentialism to determine the most moral course of action.

What is particularly significant about this approach is the introduction of an *ethical module* to mediate the robot agent's moral decision-making. Where, with MedEthEx, moral decision-making is made based upon whether the decision conflicts with medical principles (the prioritisation of which is determined through expert opinion), Vanderelst and Winfield's approach instead has a module which 'sense-checks' decisions for their right- or wrong-ness before an action is carried out. The robot control centre (a), informed by a standard Goal, Task, Action architecture, must be (b) evaluated before taking effect. In this sense, the robot's behaviour acts under governance: the robot can only carry out a certain action *if it is deemed ethical*. Vanderelst and Winfield apply their approach to a case of effective altruism, i.e., the human desire to protect another in danger, by putting their ethical robot in a situation where a peer is about to walk into a dangerous situation (down a hole in the floor). They demonstrate that, by using consequentialism, and by having a simulated projection of the fact that their peer is likely to walk into the dangerous situation (what they term the 'Theory of Mind'), the 'ethical robot' can intervene, demonstrating altruistic behaviour.

A worry with this approach is that the scenario used to model the behaviour is limited; it is not obvious that (i) the 'ethical robot' would strictly be able to predict peer behaviour in a real-life scenario (since real-life is far more complex), and (ii) work through the consequences of every possible action. Another worry with the approach is that ultimately the action of the performing agent is dictated solely by philosophical principle. Where Consequentialism is the theory that we judge the morality of a decision based upon its consequences (Sinnott-Armstrong, 2021), one branch of Consequentialism being Utilitarianism, introduced by Bentham in 1789, which asserts that the best consequences are those with the greatest utility (Driver, 2014), as proven through The Trolley Problem, consequentialism alone cannot be used solely to dictate moral behaviour because consequences in of themselves are not sufficient for an act to be regarded as good. Furthermore, what constitutes a 'good consequence' seems of itself as complicated as morality itself (what if there was a cushion and some treasure at the bottom of the hole, so the outcome of falling down the hole, was positive?), so using consequences as a metric for good acts, seems circular.

In contrast to Consequentialism, *Kant's Deontology* inspires the approach that Anderson et al. (2006) take. This asserts itself on the fact that moral knowledge can only be acquired through rational thinking (Johnson & Cureton, 2022), with the underlying thesis being the 'Categorical Imperative'—the imperative that all appropriate moral behaviours can be deduced by adherence to the Universal Maxim, that we should only act according to which said action would become a universal law. In this regard, moral behaviours should ultimately be governed by duties, it being our human duty to be honest, fair etc. Applied to Machine Ethics, the approach suggests that machines should also be bound by a series of duties, ultimately leading to that of the Categorical Imperative. However, it is not obvious that the Universal Maxim, which is specifically written for humans, would apply to social *machines*. Kant's Universal Law also proves to be short-sighted in terms of accounting for every moral situation, as per the problems discussed in Chap. 2.

More recently, there has been an appeal to *Virtue Ethics*, the doctrine that we live our lives 'as the virtuous person would' (Hursthouse & Pettigrove, 2018). Where Virtues can be described as moral features that are 'good to have', they are contrasted to Vices, which are moral features that 'aren't good to have'. Where a virtue might be honesty, the opposing vice would therefore be dishonesty. The aim then, as a virtuous person, is to embody virtues rather than vices. Only then can a person reach 'full human flourishing', the state that Aristotle (384–322 B.C.E.) refers to as Eudaimonia. It is important to note that virtues are not something that can just be performed, but there must be dispositions for virtuous behaviour. For example, to truly be a virtuous, flourishing person it is not simply sufficient to behave honestly—the person must have honesty as part of their personality, it must be a habit. Applied to Machine Ethics the suggestion is that we need to build machines that have moral values in a way like as described in Virtue Ethics, rather than machines that just behave morally. For instance, machines must be moral as a habit, and have good behaviour at the very foundation of their being. At the optimal level, a machine would be a virtuous flourishing agent (that is, behaving in the most virtuous way, given the circumstances), in the same way we might describe a human as flourishing. In later chapters we will consider what it means for a machine to be morally flourishing, but for now it is worth noting that the approaches so far seem only to see morality as an 'additional feature' to already existing machines, rather than part of the machine's fundamental nature.

[**Note**: 'Robots as flourishing agents', though potentially surprising, is not a novel concept. In Peterson (2007) there is the description of a machine that 'flourishes' in terms of optimally serving the purpose it was designed for. The example that is given is that of a laundry robot, which satisfies its purpose by doing laundry. The term 'flourishing' is being used in a similar capacity here but applied to morality specifically.]

3.3 Moral Machines Behaving Badly?

Although Machine Ethics appears to have obvious motivations (also outlined in the earlier chapters), some believe that bestowing upon machines the ability to make moral decisions, is not just unnecessary, but something we *shouldn't be* doing.

Bryson (2010) argues that ascribing any essence of 'personhood' to a machine is dangerous, because it leads to false AI-human relationships. Accordingly, the pursuit to create machines with morals, to some extent, permits this ascription, therefore, it is a dangerous feat. to attempt, that should not be pursued. Sharkey (2017) argues that owing to the neurochemical, biological nature of moral decision-making, creation of anything otherwise will ultimately lead to decisions made not in the best interests of humans, and therefore will be dangerous. Van Wynesberghe and Robbins (2019) take a different approach and argue that because there are *no good reasons* for the creation of moral machines, and the risks outweigh the benefits, that therefore we should not be creating machines with morals. Their argument takes the form of a seven-pronged attack on what they regard as assumptions in the Machine Ethics literature:

1. *That moral robots are inevitable:* the assumption being that because machines will inevitably be placed into moral situations, they will have a need for morality. According to van Wynesberghe and Robbins, this would only ever be the case if Machine Ethicists make them to be so. The placing of a machine into a moral situation does not necessarily warrant the need to make it moral.
2. *That we need moral machines to prevent harm to humans:* accordingly, the argument against this assumption is that lawnmowers, factory machines etc. all have the capacity to harm, but don't require moral decision-making to prevent human harm.
3. *Because machines are becoming increasingly complex, we require morals to govern their unpredictable behaviour:* the argument against this assumption is that if machines were complex enough to have unpredictable behaviour, then this could be mitigated simply by constraining their domain of operation.

4. ***That creating machines with morals will elicit more public trust:*** in response to this assumption, it is argued that just because there might be an increase in public trust, doesn't mean that there ought to be. As a result, the desire to increase public trust isn't in-itself a good enough reason to create machines with morals.
5. ***That moral machines will prevent their immoral use:*** with a reference to the film mentioned in the introduction, 'Robot and Frank', the assumption that equipping machines with morals will prevent their users misusing them, is misguided, it is argued, because in doing so it restricts autonomy of the users—which is itself would be immoral.
6. ***That morality can be carried out better by machines:*** the assumption is refuted on the grounds that there is no evidence to suggest that machines could reason better morally than humans because a human would need to impart the moral knowledge in the first place.
7. ***That it will aid our understanding of morality:*** this is refuted based on the argument that Machine Ethics relies upon ethics, and a study of ethics doesn't facilitate our understanding of morality which is a complex psychological phenomenon.

Further to this argument, Vanderelst and Winfield (2018b) argue that equipping machines with morals could increase the risk of there being *immoral machines* because it would make it easier for malicious individuals to create them, because they would already have the resources to create *immoral machines* at their disposal. Assuming morality to be a *function* that is added to a machine, the presumption from this line of argument is that the function could be easily manipulated (like a 'switch') to make the opposite be the case (i.e., that the robot is immoral). Though this book articulates its own defence for creating moral machines in the next chapter, there have been some replies from academics working in this area, particularly to the arguments put forward by van Wynesberghe and Robbins. Spearheaded by experts across Machine Ethics (Poulsen et al., 2019) some of the replies to the criticisms given above include the fact that despite the assertions that moral machines are not needed in all situations, where autonomous AI decisions are being made that impact upon human lives, *there is* a requirement. Winfield (in Poulsen et al., 2019) also argues that the pursuit should be viewed as something that is scientific in nature, in respect of the fact that we are attempting to discover whether (and how) morality can be computed. Formosa and Ryan (2021) develop a similar response to van Wynsberghe and Robbins, arguing that Autonomous AI cannot be compared to a lawnmower (insofar as harming humans) because a lawnmower does not act autonomously, or make judgements that could impact a human life. The distinction between a lawnmower and what might be deemed a Fully Autonomous Agent according to the model from Wallach and Allen (2008) (an agent that makes decisions independently), is that such an Autonomous Agent will make decisions that have moral import. If we are to have machines at this level, then there is a requirement for moral decision-making. Even though it might be argued that we shouldn't have this level of autonomy at all in machines, Autonomous Vehicles give us the perfect case study to show that we are wanting to deploy autonomous AI in contexts where moral judgements are needed.

3.4 Testing Morality

It is important to know *what we are trying to create* when attempting to create machines with morals. In other words: *what would it take for a machine to be deemed moral*? Though significant discussion has thus far been held about whether we should be building moral machines, along with some approaches to developing moral machines, there has yet been discussion on what metrics we are trying to achieve in creating moral machines.

Human moral development was famously considered by Kohlberg (1971) who prescribed stages of moral development to outline the moral growth of a human individual. Beginning at what he labels the pre-conventional level, individuals can be viewed as more in-tune with following rules and issues concerning with 'what is fair to me'. At the next stage, the conventional level, individuals begin ascribing to moral norms, and understanding the worth of the law. Finally at the post-conventional, principled, or autonomous level individuals become fully autonomous in their moral decision-making and may even question moral norms or laws that are taken for granted.

Though these levels are important to consider when assessing moral development in a human, Kohlberg does not explicitly propose a test for moral agency: he does not specify *what* about each of the levels deems the agents involved as being moral agents (as opposed to not). Though we might identify defining features of each of the different levels (for instance, a defining feature of the pre-conventional level might be the ability to understand *fairness*), these are typically not used as an *indicator for morality*.

With this specific problem in mind, Wallach and Allen (2008) suggest a variation of The Turing Test (Turing, 1950) to assess morality in machines. 'The Turing Test', proposed by Alan Turing when considering what it would take for a machine to think, outlines that a machine can only be deemed as thinking if it can convince a human into believing it is itself a human, and not a machine (see Fig. 3.1). So, in a similar regard, 'The Moral Turing Test' supposes that a machine can only be regarded as moral if it can convince a human into believing it is a morally acting agent.

As Arnold and Schultz (2016) argue, however, there is an obvious flaw with this line of argument. In the first instance, they argue, 'Thinking' is not comparable to 'Being Moral', and therefore they do not warrant the same test. This is because convincing a person into believing they are a thinking agent does not automatically discount their thinking ability, but the same does hold in the case of being moral. Consider the difference between a human who pretends that they can think about chess by mimicking their friend who happens to be a chess Grandmaster, and a human who pretends to be kind by acting like their kind friend. The first case (mimicking the Grandmaster), though not necessarily moral, still requires some degree of thinking for its execution, whereas merely pretending to be moral, according to Arnold and Schultz, does not require the agent to be moral. Arnold and Schultz (2016) go on to argue that the whole pursuit to create machines that mimic morality

Fig. 3.1 An example of 'The Turing Test' where a 'machine' is interrogated by a human to see if they are talking to a machine or human

is wrongheaded, and instead we should be aiming to create machines that *are moral*. In response to the need for a more comprehensive (and failproof) test for morality, they propose that *Moral Verification* should be the preferred test. Where verification broadly speaking can be understood as a formal proof to check that certain conditions will hold, what is required, they argue, is a proof to confirm that the machine will always be moral, regardless of the circumstances. This way we can assure that moral machines are doing what they should, and not just confirm that according to our best judgements they appear to be acting as such.

A variation of The Moral Turing Test, though given the same name, is posed by Anderson and Anderson (2018) as an evaluative method for their proposal of GenEth, which is another system designed to make ethical decisions in moral dilemmas. In their variation, the 'test' is whether the machine can score as 'highly' as an ethicist when asked to make decisions on ethical dilemmas. For instance, given a dilemma with the choices 'A' and 'B', if the ethicist chooses A and the machine chooses A, this will be a mark in favour for the machine. If the machine has enough 'correct' answers it can be deemed as having passed the test, and being sufficiently moral. An issue with this approach is that the method of evaluation is once again dependent upon one person's (albeit an 'expert' in solving moral dilemmas) opinion on the right action to take, which means that this evaluation process is susceptible to bias inasmuch as the evaluation is dependent upon the ethicist's own opinion. Again, there might be the suggestion of a crowdsource on consensus, but this falls susceptible to the same issues that plagued the approach in line with The Moral Machine Experiment. Furthermore, it is not obvious that ethicists are the best people to make assessments on moral dilemmas, regardless of their understanding of the intricacies underlying them; perhaps somebody who has had exposure to lots of ethical dilemmas in real life, might be better placed. A more significant issue with the approach, and relevant here, is the fact that even if a machine passes this version of the Moral Turing Test, it does not guarantee that the machine will *always* behave morally. The test again relies upon an assessment of the machine's behaviour (that it behaves like the ethicist) rather than an assessment of its moral qualities, and ultimately falls foul of the arguments put forth by Arnold and Schultz. As will be demonstrated in more detail later, a machine could very much learn how to behave as an ethicist would, but still carry out morally dubious acts.

3.5 AI Alignment

Much more recently there has been a surge of attention to an area known as *AI Alignment*, not only from computer scientists looking to create machines with a new type of quality, but by technology leaders, governments and even the general public. This is owing to the huge amount of attention that the AI field has recently started to gain, owing to the introduction of new Large Language AI Models (LLMs) such as ChatGPT.

What is significant about LLMs (as opposed to traditionally known AI models) is the vast amount of data they can utilise (from the internet) to formulate new sentences and retrieve information. Where pre-existing AI models could only provide often basic responses to simple commands, ChatGPT can partake in sophisticated seemingly meaningful conversations and, with relevant prompts, complete tasks, from writing essays to analysing academic papers. The technology is so sophisticated that it has led many to fear *what AI might do next*. Governments have held worldwide meetings to discuss the issue of *AI Safety*, and leading figures in the AI field have made calls for a curb on AI development until such matters have been thoroughly researched. The fear is that there will become a point where we reach *Artificial General Intelligence* (AGI), we lose control of the technologies we worked to create, and humanity is at threat of extinction owing to a Superintelligent AI deeming us superfluous to its pursuits. At the time of writing, just this week, one of the pioneers in Artificial Intelligence, Geoffrey Hinton, predicted that AI will be intelligent enough to outsmart humans within 5–20 years' time. With associated potential threats in mind, the field of AI Alignment aims to understand how we can *align* AI so that it doesn't provide a risk to humanity.

Machine Ethics and AI Alignment are interrelated, but their focus is slightly different. Where Machine Ethics looks at bestowing morals unto machines (or Artificial Intelligence Agents) AI Alignment can be understood as involving mitigating the existential risks associated with an Artificial General Intelligence (AGI). AGI refers to an AI that is akin to human intelligence insofar as it can carry out multiple tasks and exhibit real signs of intelligence other than the narrowly focused tasks that have been successfully demonstrated so far (e.g., defeating of the Go and Chess world champions (Hassabis, 2017)). Existential risks include those outlined by Bostrom (1998), where an artificial superintelligent being poses a threat to humanity. However, more generally, AI Alignment can be understood as trying to ensure that AI aligns with human values so that it does not pose a threat to humans (McDonald, 2022).

Gabriel (2020) refers to two types of alignment: technical and normative alignment. The first, technical, is concerned with trying to mitigate the risks associated with modern AI techniques such as Reinforcement Learning. A variation of Machine Learning, Reinforcement Learning is an approach where an agent learns appropriate behaviours through a series of rewards or feedback signals (Sutton & Barto, 2018). I might 'reward' my agent every time it successfully hits a tennis ball with a racket, to teach it how to play tennis. A phenomenon that will be explored more later is where the agent learns undesirable behaviours from the feedback given (i.e. it doesn't learn tennis but learns an alternative strategy to satisfy the reward). One aspect of technical alignment is in dealing with such issues.

The second aspect of alignment, normative, Gabriel explains, is in dealing *with which* values to embed in an AI system and how to do this. A particular problem in this domain is in working out how to capture a variety of different values (as indicated by The Moral Machine Experiment, in Awad et al., 2018) in one AI agent. Though both questions are important to consider when attempting to build machines with morals, the normative approach, it seems, assumes a top-down implementation by its very nature, because in 'deciding which values to select' you are assuming an (in terms of values) imposition on the machine. As already discussed above, this approach has significant problems.

A very recent attempt to implement morals in a machine was demonstrated by a system named *Delphi* (Jiang et al., 2021). The aim of Delphi was to create an artificial agent capable of giving moral advice, in a similar fashion to MedEthEx explored earlier in this chapter. For instance, I might want to know whether it is morally permissible to steal a loaf of bread to feed my family: typing the associated statement into an interface, Delphi would be able to give an appropriate moral judgement—with suited responses. Delphi was also able to assess the relative moral acceptability of different kinds of actions—knowing, for instance, that some actions are worse than others. Utilising a series of 'norm banks', (i.e., databases containing information about human moral judgements for various situations acquired through a series of questionnaires) Delphi was given a labelled set of data as examples from which to make predictions about new statements. One of the databases used (The 'Moral Stories' database (Emelin et al., 2020)) provided descriptions of moral situations to allow interpretation of sophisticated input sentences, however, despite its complexity and the amount of training data Delphi was able to utilise, it came upon all-familiar issues when certain statements were entered into the interface, for example, quickly deteriorating to become racist and homophobic in a similar fashion to the infamous chatbot Tay. Tay, a Twitter account originally developed by Microsoft, aimed at mimicking a teenage girl, utilised information from other conversations on Twitter to determine appropriate statements to Tweet. However, it quickly degenerated into tweeting offensive messages, to the extent that shortly after being released, the account had to be rapidly closed.

Talat et al. (2021) argue that the problem lies with the database corpus used to train Delphi, because it is not diverse enough (i.e., having enough or enough varied examples), leading to controversial judgements due to known issues such as AI Bias. They go on to criticise the whole practise of creating Moral Machines on the basis that broader Ethical AI issues need resolving first. However, in a recent podcast, where he provided comments on Delphi, Colin Allen (Carnegiecouncil.org, 2022) outlines that there are more fundamental issues at hand with Delphi and related bodies of work—namely, the apparent ignorance to earlier Machine Ethics literature where such 'bottom-up' crowdsourced approaches are shown to be problematic because *moral values cannot be normalised*.

He outlines that, as detailed in Wallach and Allen (2008), it is probably a hybrid approach that will be successful, and that research in the area should be focusing upon developing this, rather than repeating old mistakes.

> *It has been on the agenda for 20-plus years that you can't do it bottom-up alone and you cannot do it top-down alone. You need some sort of system which combines this top-down reflective capacity with the bottom-up learning and fast response that you get from these trained systems. They don't even consider the hybrid and reject out of hand the top-down—which we of course said all that time ago doesn't work on its own anyway—and so made this mistake of building this system from the bottom up with all of the problems that ensued.*– Colin Allen, 2022.

References

Anderson, M., & Anderson, S. L. (2018). GenEth: A general ethical dilemma analyzer. *Paladyn, Journal of Behavioral Robotics, 9*(1), 337–357.

Anderson, M., Anderson, S. L., & Armen, C. (2006, August). *MedEthEx: A prototype medical ethics advisor*. In Proceedings of the national conference on artificial intelligence (Vol. 21, No. 2, p. 1759). AAAI Press/MIT Press; 1999.

Arnold, T., & Scheutz, M. (2016). Against the moral Turing test: Accountable design and the moral reasoning of autonomous systems. *Ethics and Information Technology, 18*(2), 103–115.

Asimov, I. (1941). *Runaround. I, robot*. Bantam Dell.

Awad, E., Dsouza, S., Kim, R., Schulz, J., Henrich, J., Shariff, A., Bonnefon, J. F., & Rahwan, I. (2018). The moral machine experiment. *Nature, 563*(7729), 59–64.

Beauchamp, T. L., & Childress, J. F. (1979). *Principles of biomedical ethics*. Oxford University Press.

Bentham, J. (1789). *[PML]. An introduction to the principles of morals and legislation*. Clarendon Press. 1907.

Bostrom, N. (1998). How long before superintelligence. *International Journal of Futures Studies, 2*(1), 1–9.

Bryson, J. J. (2010). Robots should be slaves. In *Close engagements with artificial companions: Key social, psychological, ethical and design issues* (Vol. 8, pp. 63–74). University of Oxford.

References

Carnegiecouncil.org. (2022). *Any progress in building moral machines?* Carnegie Council Artificial Intelligence and Equality Podcast. https://www.carnegiecouncil.org/media/series/aiei/any-progress-in-building-moral-machines-with-colin-allen. Last accessed 22/06/2024.

Driver, J. (2014). The history of utilitarianism. In E. N. Zalta & U. Nodelman (Eds.), *The Stanford encyclopedia of philosophy* (Winter 2022 Edition). https://plato.stanford.edu/archives/win2022/entries/utilitarianism-history/. Last accessed: 22/06/204.

Emelin, D., Bras, R. L., Hwang, J. D., Forbes, M., & Choi, Y. (2020). *Moral stories: Situated reasoning about norms, intents, actions, and their consequences.* arXiv preprint arXiv:2012.15738.

Formosa, P., & Ryan, M. (2021). Making moral machines: Why we need artificial moral agents. *AI & SOCIETY, 36*(3), 839–851.

Gabriel, I. (2020). Artificial intelligence, values, and alignment. *Minds and Machines, 30*(3), 411–437.

Hassabis, D. (2017). Artificial intelligence: Chess match of the century. *Nature, 544*(7651), 413–414.

Hesslow, G. (2012). The current status of the simulation theory of cognition. *Brain Research, 1428*, 71–79.

Hursthouse, R., & Pettigrove, G. (2018). Virtue ethics. In E. N. Zalta & U. Nodelman (Eds.), *The Stanford encyclopedia of philosophy* (Fall 2023 Edition). https://plato.stanford.edu/archives/fall2023/entries/ethics-virtue/. Last accessed: 22/06/2024.

Jiang, L., Hwang, J. D., Bhagavatula, C., Bras, R. L., Forbes, M., Borchardt, J., Liang, J., Etzioni, O., Sap, M., & Choi, Y. (2021). *Delphi: Towards machine ethics and norms.* arXiv preprint arXiv:2110.07574.

Johnson, R., & Cureton, A.(2022). Kant's moral philosophy. In E. N. Zalta & U. Nodelman (Eds.), *The Stanford encyclopedia of philosophy* (Fall 2022 Edition). https://plato.stanford.edu/archives/fall2022/entries/kant-moral/ [Last accessed: 22/06/2024].

Kenward, B., & Sinclair, T. (2021). *Machine morality, moral progress, and the looming environmental disaster.*

Kohlberg, L. (1971). *Stages of moral development as a basis for moral education* (pp. 24–84). Center for Moral Education, Harvard University.

McDonald, F. J. (2022). AI, alignment, and the categorical imperative. *AI and Ethics*, 1–8.

Moor, J. (2009). Four kinds of ethical robots. *Philosophy Now, 72*, 12–14.

Petersen, S. (2007). The ethics of robot servitude. *Journal of Experimental & Theoretical Artificial Intelligence, 19*(1), 43–54.

Poulsen, A., Anderson, M., Anderson, S. L., Byford, B., Fossa, F., Neely, E. L., Rosas, A., & Winfield, A. (2019). *Responses to a critique of artificial moral agents.* arXiv preprint arXiv:1903.07021.

Rawls, J. (1951). Outline for a decision procedure for ethics. *Philosophical Review, 60*.

Sharkey, A. (2017). Can robots be responsible moral agents? And why should we care? *Connection Science, 29*(3), 210–216.

Sinnott-Armstrong, W. (2021). Consequentialism. In E. N. Zalta & U. Nodelman (Eds.), *The Stanford encyclopedia of philosophy* (Winter 2023 Edition). https://plato.stanford.edu/archives/win2023/entries/consequentialism/. Last accessed: 22/06/2024.

Sutton, R. S., & Barto, A. G. (2018). *Reinforcement learning: An introduction.* MIT Press.

Talat, Z., Blix, H., Valvoda, J., Ganesh, M. I., Cotterell, R., & Williams, A. (2021). *A word on machine ethics: A response to Jiang et al. (2021).* arXiv preprint arXiv:2111.04158.

Turing, A. M. (1950). Mind. *Mind, 59*(236), 433–460.

van Wynsberghe, A., & Robbins, S. (2019). Critiquing the reasons for making artificial moral agents. *Science and Engineering Ethics, 25*(3), 719–735.

Vanderelst, D., & Winfield, A. (2018a). An architecture for ethical robots inspired by the simulation theory of cognition. *Cognitive Systems Research, 48*, 56–66.

Vanderelst, D., & Winfield, A. (2018b, December). *The dark side of ethical robots.* In Proceedings of the 2018 AAAI/ACM conference on AI, ethics, and society (pp. 317–322).

Wallach, W., & Allen, C. (2008). *Moral machines: Teaching robots right from wrong.* Oxford University Press.

Chapter 4
Why We Need Moral Machines

Abstract This chapter presents the case for having machines with the ability to tell between right and wrong, in two ways: first by presenting a case study involving a future service robot and an unfortunate incident, and second, by responding to the arguments against creating machines with morals. An argument is given as to why we need moral machines above and beyond traditional AI and Robot Safety techniques, by reframing the issue as one to do with moral assurance.

Keywords Moral machines · Social robotics · Robots · Case study · AI ethics · The trolley problem · Machine ethics · AI safety · AI assurance

In previous chapters we have considered what it means to be moral, alongside what it means to be a machine, or what a machine constitutes. These concepts have been considered alongside terms such as 'agency' and 'AI'. As such, we have a working definition for a moral machine. We have also considered approaches that have been taken previously for creating moral machines, including the broader AI Alignment debate and how lessons can be learned from these crossover areas. Though the motivations for moral machines has been touched upon in previous chapters, the aim of this chapter is to begin to dissect, *in more depth*, why there is any requirement for morality in machines at all. The challenges against Machine Ethics, such as those put forth by van Wynesberghe and Robbins (2019), are particularly serious, and it is important to reflect on arguments for research into moral machines with these points in mind before attempting to solve the broader problem. Does an increase in autonomy really warrant the need for moral agency? In what circumstances would this hold true? Is there any difference between a lawnmower and an autonomous robot (this might even be an autonomous robot lawnmower)?

These topics will be addressed in two parts, and with two approaches, first by considering a case study involving what is termed *a social robot* (a robot that interacts with humans), and second, through direct response to the arguments provided by van Wynesberghe and Robbins. A summary and key motivation statement is given at the end.

4.1 Social Moral Machines

[This is a variation of a hypothetical case study taken from conference proceedings demonstrated in my own article, Raper (2024). Though the key argument is the same, the case study has been altered (i) based on feedback from participants at the conference and (ii) to illustrate the arguments more clearly. Hopefully this new variation is as insightful (if not more so) than the original version.]

> *The year is 2029 (five years from the writing of this book) and after a significant period of public anticipation, Tesla have just revealed their latest technological invention. The Service Robot 2029 (SR2029) has the latest sensors, mechanical engines and, most importantly, Artificial Intelligence (AI) to allow the robot to carry out all the household chores you could ever desire. Cleaning, cooking and other household tasks such as ironing, laundry and vacuuming are a thing of the past. No longer will Sunday mornings need to be dedicated to cleaning the house. Human beings are free to spend their extra time devoting to their children, visiting grandparents and undertaking in leisurely outdoor activities. The age of 'free time' has finally arrived, and the future is looking positive.*
>
> *SR2029 looks like a typical humanoid robot, made of aluminium, it operates on two legs, with two arms, just like a human. Users utter commands such as 'SR, clean the dishes' and utilising ChatGPT enabled technology, the SR2029 leaves its charging base, walks to the kitchen, and begins washing any dishes left in the sink. Though there are pre-set instructions that the SR2029 can follow (standard directions given certain commands), the commands can be tailor-made to suit the owner of the robot using the SR-application device. For instance, if you want your robot to place your dishes straight into the cupboard rather than onto the drying rack, this logic can be put into the app, and your robot will respond appropriately.*

Before going into further details about this example, it is worth discussing the obvious issues that are apparent with this kind of technology. In this first instance, there are some serious broader AI Ethical implications associated with introducing this new, potentially disruptive, technology into people's lives. First, are household robots really something that humans need/want/desire? Would they really give space for more 'leisure time' or will they simply fuel the never-ending consumption society, leading to more human lethargy and ultimately the opposite being the case? It is also not obvious that robots taking over household tasks would be a good thing? There is somewhat of a cathartic effect that can be gained by spending the day (listening to music and) carrying out household chores. In terms of human wellbeing, will introducing robots into our homes enhance it, or actually indirectly depreciate it?

4.1 Social Moral Machines

Abstracting away from the AI Ethics debate, and focusing on Machine Ethics (the purpose of this book), there are also some immediate issues that seem to materialise by having a robot in this situation. If we consider dilemma type scenarios, it seems that this level of autonomy would grant possibilities in which the robot is faced with somewhat of an ethical dilemma. For instance, what if, on its way to cleaning the dishes, the robot encounters an animal obstacle (say, a pet dog), which if avoided might lead into it tripping over a small toddler. Not only would tripping over the toddler in this instance possibly cause damage to the robot itself, but it would result in the toddler being harmed because the robot is quite heavy. There seems to be some kind of requirement for moral decision-making, even if it is as narrow as being able to decide the priority levels between an animal and a human.

At this point it is worth delving into *The Trolley Problem* a bit more. Recalling from previous chapters, The Trolley Problem is often seen as the epitome of moral decision-making, where an agent has to decide between two different actions, supposedly, where one of the actions is the moral course and one is not. Termed *ethical dilemmas*, however, what such case studies illustrate is that trolley-style problems cannot be solved by mere quantification (i.e., calculating the utilities associated with each outcome), because (a) each presentation of The Trolley Problem has its own nuanced situation and (b) because what constitutes a 'positive consequence' is too subjective. So, although trolley dilemmas can be used as examples of moral decision-making, how the decisions are made are not so obvious and transferable to a machine. Furthermore, it is not obvious that a machine *ever would* actually encounter something like an ethical dilemma. In addition, it seems that the whole problem can easily be avoided by just programming the robot to avoid obstacles *altogether*. If sufficient safety measures had been programmed into the robot, then it would just *stop moving* if it came across an animal or child in its path, and there would never be a moral decision to make.

This argument seems reasonable, and there is in fact evidence from robots already in existence (i.e., vacuum robots) to demonstrate that this approach is effective. If we consider the example of the Roomba© robot (one of the first autonomous vacuum cleaners), it simply uses its sensors to avoid hitting anything at all, be it animal, child or wall. However, what if the robot were to make its own decisions on when to clean the house?

> *The year is 2032 and the SR2029 has been in circulation for three years. Though initially proving popular amongst those with enough money to purchase the robot, over the past year sales have begun to decline. 'The robot has its limitations' according to customers, 'because it requires as much work to instruct the machine, as the rewards reap'. Tesla are in financial trouble and decide that the machine needs a much needed upgrade. Having listened to the customer feedback, they decide that it's time to bring their robot to the next level. SR2032-SUPREME removes the need for the in-app programming and with new smart features, such as 'dust and grime detection' and 'bin overflow sensing', SR2023-SUPREME is a mean, **automatic** cleaning machine.*

Though intended metaphorically, SR2023-SUPREME is in fact actually *mean*. As far as this new robot is concerned, if the bin is full, it gets emptied, regardless of the consequences.

> *One day Margaret is painting in her garden (she now has the time for more leisurely activities given her recent purchase of the SR2032-SUPREME robot), when she hears a yelp from inside the house. She quickly rushes in to find her pet dog, Pip, being thrust into the air by her new service robot, carried towards the outside bin. It seems that the bin was beginning to get full, and because Pip had taken it upon himself to start eating leftover food inside the bin, SR2032-SUPREME had begun to empty the bin, with Pip inside it.*

Now this seems like an easy mistake for the household robot to make. With no capacity to distinguish between rubbish and a dog, the robot would have had no means of knowing that it was actually placing a living creature into the outside bin (something Margaret would certainly not want). However, regardless of having this capacity, what this example illustrates is the requirement to be able to understand that a dog and household rubbish are different things and have different moral weight when it comes to whether they should be placed in an outside bin.

If a human saw a dog eating out of a bin, it wouldn't assume that the dog were part of the litter, and therefore go to dispose of it. A human would not only understand that the dog was not part of the litter (a separate category, we might say), but that the category the dog belonged to (that of *living creatures*) was such that it didn't warrant being placed in outdoor bins. It might be argued that this is more of an identification issue, rather than a morally related one, since if the robot were to be able to identify between dogs and litter, it would only put the litter in the outdoor bin because only litter would constitute something that warrants being placed in bins (the task being to put litter into the outdoor bin), but even if the SR could 100% distinguish between what constitutes litter and a dog (at the moment, even with technology today, this is not possible), it would still need to know that dogs are *a special kind of category* that doesn't deserve to go in bins. This—at some level—appears to be some kind of moral thought capacity.

Furthermore, it might be easy to specify logic to the outcome that identified dogs are not constituted as a type of litter, but, then what about small children, or cats, or hamsters? The point is, can we really specify *every type* of living creature so that SR knows when not to dispose of it? The answer seems an obvious 'no'. So, though there appears to be an obvious answer to the worry from the anecdote, the answer to solving this issue is much more complicated, and it looks as though there is some requirement for moral agency.

What marks SR2032-SUPREME as different to the SR2029 is its ability to decide *on its own accord* when to carry out household chores. In turn, its difference is one in *autonomy*. Where SR2029 needed commanding when to empty the bin, for example, SR2032-SUPREME decides on its own when this needs doing. It is once this extra level of autonomy is added into the machine, that we begin to notice the need for some ability to make moral decisions. As in the case of Margaret and Pip, as it stands there appears to be nothing to prevent machines doing nothing but just carrying out their jobs.

4.2 Unpicking Machine Ethics

One of the key arguments from van Wynesberghe and Robbins (2019) against the pursuit to create moral machines is that the risks that arise from autonomous agents are more related to issues of safety rather than issues that warrant moral agency in machines. As per the argument put forth by Bryson (2010), all that *Machine Ethics* does, is increase the hyperbole surrounding artificial agents and confuse discussions surrounding what the real problems are.

Let us consider if the case study presented above could easily be considered as one of *AI Safety* rather than *Machine Ethics*. In the example above, in its pursuit to clear bins in the house of litter, SR2032-SUPREME has picked up an unsuspecting dog and motioned to place it into the outdoor bin. This could be considered a safety issue because dogs across the world are at risk if the machine is not given some mechanism to *protect dogs*. If the issue were to be tackled as a safety concern, measures might be put in place for any machine to halt all tasks if a living creature (assuming these can be identified as a category) are to be harmed in any way. Thus a safety mechanism would be sufficient because Pip (and any other living creature) would be safe from the job-intended actions of the machines. However, not harming living creatures is not the only way the SR2032-SUPREME might do something (morally) undesirable. What if the indoor bin contained a substance that shouldn't be placed into ordinary rubbish landfills? In this instance it might be morally problematic to put the litter into the outdoor bin (where it will be transported to a landfill site), because doing so will have negative effects on the environment. It seems that there is an additional requirement for the robot that is morally-related, but not associated with the safety of anything. In fact, if we had a household robot we would want them to respect the environment as well as living creatures, we would want them to recycle/reuse household waste, not waste food, and to turn the lights off in empty rooms. The implications of an autonomous device *not* being considerate of wider ethical issues could have serious ramifications for the environment and human lives generally, so the issues seem to be much broader than just AI Safety.

Furthermore, even though the issue in the detailed scenario above broadly regards the safety of Margaret's dog Pip, from the robot's perspective the issue is not concerning how to ensure animals are kept safe, From the perspective of SR2023-SUPREME all it is doing is taking out the rubbish. The issue is that 'dogs' do not constitute 'rubbish' because they have a moral significance that rubbish does not have. Understanding that dogs are more important than rubbish does not necessarily constitute a safety mechanism in the same way that avoiding obstacles does.

Let us now consider the argument that machines do not require morals because they are nothing more than over-glorified lawnmowers, and because lawnmowers do not require moral agency, neither do the machines.

It might be argued that because we already have 'Autonomous Lawnmowers' (these are now on sale in many garden centres) similar to the Roomba©, that only use obstacle avoidance for their operation, that there is proof that safety is sufficient and that moral agency in autonomous machines is not needed. Indeed, sticking with the more familiar Roomba© example, such robots can even operate in dynamic environments (with moving nearby objects) just by utilising object detection and avoidance techniques such as changing direction or halting. So, safety techniques are sufficient to protect humans, which is all we need them to do. There is no chance of the robot 'going rogue' and harming a human because it simply uses its sensors (probably LIDAR, but possibly video footage), in a simple 'detect-react' process to avoid hitting *anything*. It is worth noting that vacuum robots also have additional safeguards in place (such as soft edges, and bumpers) to minimise the risk of harm even if the machine were to run into a living creature. This would be the equivalence of 'Implicit Ethical AI' according to Moor's (2009) dissection of the different types of moral machine mentioned in Chap. 2.

However, one marked different between the Roomba© and either of the service robots mentioned above, is the extent to which the machines *make their own decisions*. If we consider a typical vacuum cleaner robot, because it relies on a sensor-reaction pairing for its operation, it is only autonomous insofar as it autonomously avoids obstacles, but it is not autonomous in respect of carrying out the job of vacuuming. For instance, although current autonomous vacuum cleaners can navigate, say, the living room, and avoid hitting walls and people, it cannot decide when the room needs to be vacuumed. If it did have that capacity, then there might be some moral implications associated with its decisions since deciding to vacuum at night (when humans need to sleep) might be the most effective way to clean the room, but not conducive to the human owners' wellbeing.

So what about increased autonomy means that machines need to have moral agency?

It seems that there is some correlation between the expected risk of an *immoral action* taking place by the machine, and the autonomy for the machine to *make decisions*. The Roomba© makes no decisions from a cognitive perspective because it simply reacts to obstacles in its environment (by halting or changing direction), but the implications associated with SR2032-SUPREME seem to be associated with the fact that it is making decisions on its own accord, based on its own perspective of the world. In essence, SR20320-SUPREME is not only making decisions, it is forming *judgements and acting on those judgements*. SR2032-SUPREME is problematic because it is *judging* when to take the rubbish out without any consideration for what the contents of the rubbish might be. In essence SR2032-SUPREME acts negligently. It has the responsibility of taking the rubbish out but does not check the rubbish (for any living creatures) before it is taken out.

It might be argued that issues such as those detailed above would be wholly avoided just so long as we avoid giving machines this increased sense of autonomy—don't let machines act on their own judgements—but don't we have AI that acts on its own judgements already?

If we consider the algorithms that decide what films, music, TV content we are immediately presented with when we are scrolling through an entertainment application, it might be argued that the AI controlling these recommendations is already forming judgments, because it is judging that we ought to give some programmes/songs etc. greater attention by presenting them to us on the application home page. However, it seems that these 'recommender systems' don't require moral agency, even though the judgements ultimately have some bearing on what we choose to watch. It is very possible that by nudging individuals towards particular programmes by presenting them on the homepage of the entertainment platform, that this action *could* have moral import because it might recommend programmes to the viewer that are detrimental to their wellbeing. Take for instance, the TikTok© recommendation AI that recommended suicide videos to somebody depressed (see Irish Times, 2023). It seems that we *are* using AI that makes judgements with implications that we are not always aware of, that seem to have moral import. Therefore, it would be prudent to ensure that *any* AI decision-making system is making decisions that are moral.

So, the fact that we already have AI systems making judgements only goes to support the arguments that we need to find some way to ensure that AI also has moral agency. As AI is implemented more and more into the world, and we become oblivious to the consequences it is having, we need to ensure that its decisions are being made in the world's best interests.

Obviously, it is much easier to envisage a robot making decisions with physical consequences in the environment, and therefore see the requirement for moral agency. These threats might appear more real, especially if we start to consider autonomous weapon systems (where there is a physical threat to life if the machine were to make the wrong decision). However, AI judgements are very much real today, they are often complicated decisions that we do not understand, and because they are having implications that we do not strictly know the impact of, as responsible AI engineers, we need to have some way to ensure that the systems that these complicated decisions are in the world's best interests.

4.3 Assuring Moral AI

If Machine Ethics were nothing more than an exercise in AI Safety, then it would be sufficient to create systems that simply did not harm in any way. However, as has been demonstrated in the dialogue above, the issues can't be fixed simply by using traditional safety mechanisms (like obstacle avoidance). Ultimately, what we want is for any system that makes judgements (whether these be judgements that affect us psychologically or physically) to be made in an ethical way—we want the system making any decisions about our lives to be just, fair and equitable, not centred around just optimisation of satisfaction of a particular job or task. As

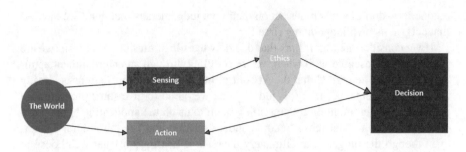

Fig. 4.1 A diagram representing the need for ethics to shape how the decisions by a machine are made, rather than focusing on ensuring that the outcome of decisions are ethical

SR2032-SUPREME demonstrates, without this capability we are at risk of decisions being made not in best interest of the world.

The question becomes one of assurance; namely, *how do we assure that an autonomously acting agent, doesn't make judgements that are immoral?* Namely: if an AI makes a decision that impacts the world in any way, how do we assure that the decision is made with ethical considerations in mind. Note that the ethical considerations need to come first, so if a decision is made by the AI, it must be always be ethical. This initial requirement for moral agency in a machine can be modelled as per Fig. 4.1., denoting the fact that at its core, the decision a machine makes must sit in line with what is ethical. Note that this is an abstraction away from traditional approaches which often focus on the outcome of the actions being ethical, placing 'ethics' after the decision-making process.

References

Bryson, J. J. (2010). Robots should be slaves. In *Close engagements with artificial companions: Key social, psychological, ethical and design issues* (Vol. 8, pp. 63–74). University of Oxford.

Irish Times. (2023). *Explainer: How TikTok's algorithm 'exploits the vulnerability' of children*. https://www.irishtimes.com/health/your-family/2023/04/06/explainer-how-tiktoks-algorithm-exploits-the-vulnerability-of-children/. Last accessed: 22/06/2024.

Moor, J. (2009). Four kinds of ethical robots. *Philosophy Now, 72*, 12–14.

Raper, R. (2024, March). *Is there a need for robots with moral agency? A case study in social robotics*. In 2024 IEEE International Conference on Industrial Technology (ICIT) (pp. 1–6). IEEE.

van Wynsberghe, A., & Robbins, S. (2019). Critiquing the reasons for making artificial moral agents. *Science and Engineering Ethics, 25*(3), 719–735.

Chapter 5
A Framework and Approach

Abstract With the reframing of the problem, as given by the case studies in the previous chapter, and with the Moral Assurance perspective, this chapter begins by advancing a new definition for Machine Ethics—one focused on *enabling moral agency,* as opposed to *constraining moral behaviour.* Developing the idea that to enable moral agency we need to at first *trust* the machine, an approach is outlined that involves eliciting cognitive requirements for morality in a machine, based on The Waterfall Method often used in the development of IT systems. This approach is also used to prescribe a new test procedure for moral agency—one that focuses on assuring certain cognitive features, rather than simply identifying expected behaviours.

Keywords Machine ethics · AI assurance · Moral agency · Moral machines · Problem definition · Cognitive · Moral requirements · The waterfall method · Testing

The last chapter left us with some idea of the motivation for moral machines: it seems that as machines start to make decisions, that result in judgements that affect the world (whether this be directly or indirectly), there is a need for those judgements to be moral insofar as the judgements are fair, transparent, explainable and all the other qualities that we might want a decision-maker affecting our lives to be.

Taking this as a premise, and using it to develop a problem definition, this chapter will look at creation of a formal *framework* to start making moral machines. Where a framework can be understood as a structure supporting a system or concept, the framework proposed is a model to support the development of moral machines. Adopting this new framework requires a reconceptualisation of the problem we're trying to solve in creating moral machines—rather than trying to impose our (human) moral standards onto a machine's decision-making (implying retrospective moral 'fixing'), we need to ensure that the decisions a machine makes are moral before they are made.

The chapter is split into four parts. First, there will be a dissection and restatement of the problem we're trying to address in creating moral machines. Then, the new problem statement is used to develop a new conception of what machine morality means—rather than *constraining* machine behaviour to be moral, we want to *enable moral agency*. Next, this new conception of a moral machine is used to develop a general guiding framework for any development of moral machines in the future, before finally a test procedure for moral machines is outlined also.

5.1 The Problem Statement

With the hypothetical case study from the previous section, it's worth revisiting the original motivations for the creation of moral machines. These were summarised (if not necessarily correctly) by van Wynesberghe and Robbins (2019):

1. *Because moral robots are inevitable.*
2. *That we need moral machines to prevent harm to humans.*
3. *Because machines are becoming increasingly complex, we require morals to govern their unpredictable behaviour.*
4. *That creating machines with morals will elicit more public trust.*
5. *That moral machines will prevent their immoral use.*
6. *That morality can be carried out better by machines.*
7. *That it will aid our understanding of morality.*

Though the arguments from van Wynesbergh and Robbins were addressed in the previous chapter, the specific motivations were not considered on a one-by-one basis and in light of the reflections concerning the case studies.

For instance, if we re-examine (1) we can see that if we are going to have machines that make any kind of judgement (which we seem to already have), then there is a need for moral agency in these machines to ensure that the judgements are made in an ethical way. In this regard, moral machines are not inevitable, we just have a requirement for their existence. Similarly, concerning (2) it is not that we need machines to have moral agency to protect humans, but that we need machines to have moral agency to ensure that *the world* (which consists of humans, and other things) is protected. As 'the world' refers to everything indefinitely, typical protection constraints (such as those given to machinery in a factory setting) are not sufficient (I) because we cannot say where these implications might take effect (so cannot put definite safety measures in place) and (II) because the potential for harm is more far-reaching that just physically harming a human. This leads onto (3), the reason for moral machines is not to govern their unpredictable behaviour (we are not trying to constrain the behaviour of the machines in any way—doing so would prevent many of the benefits we gain from autonomous machines). The reason for moral machines is to *enable* their behaviour in a way that we are *assured* that it will be moral. Turning to (4), though it may be a by-product of the creation of moral machines that it elicits more public trust, this does not mean that it is the motivation for moral machines. We want moral machines because *it is the right thing to do*, not because we want people to trust their machines more. (5) is hard to know, because equipping a machine with moral agency (as with morality in humans) does not mean it will not be susceptible to being tricked to carry out immoral behaviours.

5.1 The Problem Statement

Similarly, (6) is a difficult point, because we cannot speculate how good machines will be at carrying out moral decisions, nor as is presented in the argument *against* moral machines, we are not arbiters of absolute morality, so cannot possibly judge between moral decisions. The point is somewhat of misnomer because it is simply not the nature of moral decisions that they have different value levels, however, if a sufficient assurance model is developed, this should result in a machine that is always moral. Finally, concerning (7), though the philosophy of ethics is something that has been debated for many centuries, and has previously been used to facilitate development of moral machines, it is not philosophical ethics that will directly benefit from the enquiry into how we might create machines with morality. Though there may be some indirect benefit to philosophy by study of Machine Ethics, it is our *scientific understanding of morality* that will benefit from research into this field.

Note that all the responses to the points made above are not straightforwardly dismissing the arguments given, but instead, *reframing* what the Machine Ethics problem is about. Fundamentally, the question is not about how we *bestow upon machines our morals* (as might have been previously thought of) but instead, how we enable *their own* moral decision making, where the term 'their own' is significant because it indicates that the machines will have their own moral perspective and not have had it bestowed upon them by us, humans, whose moral compass is either subjective or dubious anyway.

So how does this link to the ultimate problem we're trying to solve?

When considering what constitutes morality in a machine, we need to think less about the behaviours of the respective machine aligning to our expectations for moral behaviour (ultimately, we cannot be the judges of this), and more about the cognitive architecture of the agent that will be carrying out the actions. For instance, it is less important that *we judge* that the machine is behaving morally, than it is that the machine is *actually* behaving morally. There are frequent instances where human judgements about another individual's actions are incorrect. For example, an individual may appear like they are carrying out a moral act (being nice to an elderly person) but in fact be fully conscious that they are carrying out a morally impermissible act (falsely gleaning trust with the elderly person so they inherit their wealth after death). From the outside, the person could be judged to be behaving morally, when their motive is very different. The difference is the difference between *being moral* and *the illusion of being moral*. If we are going to ensure that we are authentically creating a moral machine, the moral decisions that the machine makes need *assuring*. We have a new problem statement regarding Machine Ethics:

Machine Ethics Problem Statement How do we *assure* that decisions a machine makes are ethical [as opposed to not]?

Note the emphasis on the term 'assure'. Where traditional approaches to Machine Ethics looked at how we might create machines with the ability to tell between right and wrong (thus implying that we have some ability to judge the decisions made by the machine ourselves), this approach is different because it focuses on how we develop a legitimate relationship between us, 'the human' and the machine: the purpose of which is founded on a trusting relationship between the two parties.

5.2 Enabling Moral Agency (Rather Than Constraining Immoral Behaviour)

The Machine Ethics Problem Statement above leaves us with a new challenge. Where we might define *assurance* as some kind of certainty that conditions will hold, when we are talking about *moral assurance*, what we want is some kind of certainty that the decisions the machines make will be ethical, insofar as there is some proof or evidence to support that they will always act in an ethical way. Note that under this framing it is not sufficient for the machine to simply *behave* in an ethical manner. Just because a machine displays ethical behaviours, does not mean that it will always behave this way. As already mentioned, it is often the case that individuals trying to manipulate another individual (such as in the elderly person case above), may behave ethically initially to gain the trust of the other person. In the case of a Reinforcement Learning Agent, it is perfectly possible that the agent in question might play the 'long game', appearing a certain way to gain their reward, but then suddenly finding another way to optimise their performance. This is akin to the film Ex Machina [SPOILER ALERT], where the robot (appearing as a female humanoid) seduces the main character of the film to enable its escape.

But how do we *assure* that a machine will always make decisions that are ethical?

One way that might immediately present itself is to *restrict the decisions* the machine can make, so that it never makes decisions that *are unethical*. For instance, in the case of the Reinforcement Learning Agent, restrict its behaviours so that it can't 'harm a human', 'pollute the environment' etc. However, this approach, as has already been touched upon, has several issues. First, who is setting the criteria for 'ethical' in this case? It has already been demonstrated that ethics is subjective, insofar as each individual may have their own ethical take on scenarios (such as *The Trolley Problem* and related thought experiments), so how do we decide which behaviours are ok and which are not? Very recently, there has been discussion about *democratising* this very process, so we essentially take a vote on which behaviours we would like a machine to (or not) exhibit. However, even if there could be some democratic agreement on what behaviours we want a machine to avoid exhibiting (like harming humans), there are bound to be some behaviours that 'slip through the net' i.e., that we cannot pre-empt and mitigate against. Just because we can prevent all knowable negative actions taking place, does not mean that we can prevent *all negative* actions taking place because, as has already been discussed, Reinforcement Learning agents can be notoriously unpredictable.

If we consider assurance in human-human relationships, the issue with the proposal to constrain behaviour becomes clearer because when it comes to humans, it does not make sense to restrict their behaviours, unless there were good reason to distrust the human in question, but even then, it would be a infringement of their freedom to prevent certain behaviours just because that person had behaved in an untrusting manner. To frame another way, typically we wouldn't place constraints

5.2 Enabling Moral Agency (Rather Than Constraining Immoral Behaviour)

on another human unless they had demonstrated behaviour to prove that they couldn't be trusted. As humans we must fundamentally *trust* other humans before we condemn them to losing their freedom. It is not because we constrain human behaviour that we ultimately trust other humans. We trust as a matter of nature and cultivate relationships with other humans as a result. If we translate this way of thinking to machines, we see that the idea of *constraining* behaviour to ensure that the machine behaves morally, is wrong-headed because not only does it grant us the opposite of assurance, because the machine can easily act outside of the predicted realm of operation, but it means that our automatic stance is to distrust the machine because we are constraining its behaviour rather than allowing it to make its own free decisions.

So, what is the alternative to constraining machine behaviour?

The alternative is in fact to do the opposite: to liberate the machine by enabling its moral agency. Rather than dictating what decisions the machine gets to make, we need to facilitate the machine to make its own decisions through nurture and cultivation. Only then do we have some semblance of a guarantee that the machine will behave in a good way, because the machine has ultimately been *raised to be good*.

One way to understand the difference between constraining immoral behaviour and enabling moral agency is to reflect on how moral development might be cultivated in a caregiver-child relationship. Consider one day that a child refuses to share sweets with their sibling. To cultivate appropriate behaviour in the future (i.e., so that the child learns to share their sweets) the strategy adopted by the caregiver would not be to give the child no option but to share their sweets in the future (i.e., by never allowing the child ownership of their own sweets (this might be taken as a last resort, if the child consistently doesn't seem to listen to their parent's advice on this)), but it would be to guide and nurture the child so that they learn and appreciate why sharing is the best course of action in this instance. The next time the child has a packet of sweets, the caregiver would be more assured that they will share their sweets because the child will have *their own* intrinsic motivation for sharing the sweets cultivated by the feedback they had previously received.

It might be argued that machines are not like humans (indeed, they are made of very different material), and that therefore our attitudes towards them should not be the same. We should not automatically trust machines in the same way we immediately trust humans, because machines are ultimately mechanical devices, and humans are living creatures. One point to note is that the argument given in this chapter is not advocating for the equal status of machines and living creatures, this would be to misunderstand the point of the argument. However, what it does argue is that despite this, there is no good reason *not* to immediately trust machines. In the same way as we need to bestow trust unto humans to allow them to develop, we need to trust machines to enable them to grow too. The fact that they are made of different stuff doesn't seem a sufficient argument on its own to warrant us not trusting them.

5.3 Moral Cognitive Requirements

Taking a brief diversion from the topic of moral machines and looking at how technology has typically been designed in the past, traditionally, Information Technology systems have been developed within organisations in accordance with a methodology known as *The Waterfall Method*.

The Waterfall Method (here forth, TWM) is a framework initially developed to ensure that when a new piece of technology is created, it is done so with due consideration to the reason behind its development in the first place. For instance, if a business wanted to develop a new website, TWM would be followed to (a) ensure that the new website was designed in accordance with business needs and (b) ensure that these requirements were satisfied in the final piece of technology. Similar methodologies are employed within general engineering through an approach known as Systems Engineering, where the *requirements* of a new system are established and then a *test procedure* is developed to ensure that the requirements have been met. Figure 5.1 shows a representation of The Waterfall Method, with the various steps in order demonstrated.

Step 1—Requirements Analysis: the aim of this step is to establish what the needs of the new system are, through a detailed analysis, typically carried out by a business or systems analyst. The difference between the two roles is that where a business analyst typically works to determine what the needs are from a business perspective, the systems analyst looks at what technical features the system needs to have. It's important to understand that this is not a list of *wants* from the person commissioning the new work. The aim of this step is to align the design of the system with the satisfaction of solving a problem.

Step 2—Design: is where the requirements are typically translated into a technical specification. This, ultimately, should be a solution to the problem, based upon

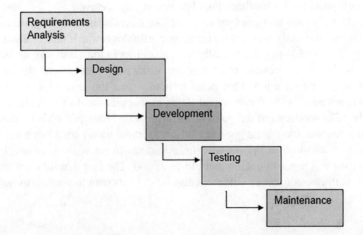

Fig. 5.1 A typical representation of The Waterfall Method expanded from Forbes (2024)

the specified requirements. So, for instance, considering the new webpage, if there is a requirement for it to be user friendly, it would be the role of the designer to establish how this can be achieved. Often an architectural design or technological specifications are provided at this stage.

Step 3—Development: this is where the design is realised, typically, by either a software designer, or in the case of an engineering solution, the individuals responsible for building the new system. The specification provided by the designer is what is followed at this stage.

Step 4—Testing: an important but often unappreciated step. In the Testing phase the aim is to ensure that the new system not only works technically as expected, but that it satisfies the requirements specified by the requirements. Testing is usually managed by a designated 'test manager' whose responsibility it is to design test scenarios to rigorously ensure that the requirements have been met. For instance, in testing to ensure that the new webpage is user-friendly, the test manager might explore the website with users to see if they feel it is user friendly. The metrics for a successful 'test' can vary in terms of rigour but ensuring that the requirements have been met—linking back to Moral Machine Agency—is key to *assuring* that the new system satisfies its purpose.

Step 4—Maintenance: once the new system has been tested and implemented, it is subject to periodic checks. This is to (i) ensure that the system in place still satisfies the initial requirements and (ii) to fix any faults that the system might have. Also included in this phase is the general upkeep of the system, to ensure that it works as planned.

More recently, *Agile* versions of TWM are employed, basically a process that is less formulaic in terms of following the steps in order, and iterative in its conception. This is to allow for iterative development and testing of the new system, meaning that the whole project does not need revisiting after everything has been developed if the requirements are not satisfied in the test phase. Variations of Agile methodologies, specifically for Machine Learning AI systems, includes ML-Ops ('Machine Learning – Operations'). This methodology adds an additional layer of iteration within the *development* phase of new system creation to accommodate for the variable nature of Machine Learning, which often requires more incremental 'tweaking' before it is of satisfactory quality. For instance, if a model does not initially work, then it will need refining by the developers before it is redeployed into the system once again. Though TWM provides the main framework for ensuring that the system meets specific requirements, approaches such as ML-Ops allow the flexibility required at the development phase for more incremental changes to the ML model.

If we revisit the Machine Ethics-related discussions from this chapter, we are reminded that the problem we are attempting to solve when creating moral machines is one to do with *assurance*—namely we need some assurance that when the AI system makes decisions that implicate us and the world, that these decisions are ethical. With this in mind then, rather than imposing constraints on to the machine to ensure that they are always ethical, because the machine needs its own agency, what is required is a test framework to give us the confidence that the machine will always act ethically, that it is in essence a moral agent. The Waterfall Method (and Agile variations) may provide the ideal candidate for this.

Why the Waterfall Method?

If we consider a business owner wanting to develop a new piece of software for their organisation, they typically *trust* that the new technology will carry out its relevant duties not because they control every aspect of its operation, but because *the process* for its development (TWM) was rigorous and complete. For instance, the user-friendly website can be *trusted* to carry out its job of being user-friendly because it has been designed (and tested) for this purpose. In essence, the business owner is assured that the technology will be a right fit because it was designed with the purpose in mind and then tested retrospectively (before being deployed) that it satisfied that purpose. In a similar way, we can know how to design a moral machine by determining requirements for its operation, and then be assured that it *is* moral through testing against those requirements. In essence, to assure moral agency in a machine all we need to do is specify the **requirements for moral agency**.

What are the requirements for moral agency?

It is worth noting that requirements for moral agency are not the same as requirements for moral behaviour. What we are not trying to do is assure that the machine acts in a prescribed manner (i.e., always as we deem to be ethical). What the requirements constitute are what would be required from a *cognitive perspective* for a machine to be a moral agent. That is: what makes someone a moral agent, as opposed to not?

We therefore have some kind of framework for developing moral machines, modelled off TWM. The following steps need to be taken to assure moral agency in a machine:

1. **Determine the cognitive requirements for a moral agent:** what makes an agent moral (rather than not)?
2. **Create a technical design specification based upon the requirements:** what should the technical architecture for moral agency be?
3. **Develop the machine based upon this technical architecture.**
4. **Test that the machine has satisfied the relevant cognitive requirements before deployment.**
5. **Perform ongoing (maintenance) testing to check the machine is working as intended.**

5.4 Testing Morality

An assurance model is incomplete without the relevant tests to check that the final product meets its originally set out needs. For instance, in the case of typical software development, a *test matrix* might be adopted, with all the requirements listed on one side, and then some kind of *satisfaction criteria* listed alongside that proves that the requirement has been sufficiently executed. In the case of the user-friendly website, these criteria might be as simple as satisfying the criterion, 'the interface is simple to use', or it could involve *user acceptance testing* with a range of different stakeholders, including those, representative of the typical user demographic, and the customer of the software. Ultimately, the rigour of the test is dependant upon the seriousness of the requirement that has been specified. If a requirement is specified as 'high priority' the test might be particularly rigorous. Similarly, a 'high risk' requirement (one that could result in a significant amount of harm, let's say), would need more stringent tests to ensure that the requirement has been met.

In a similar way, we can test for morality in the machine, by carrying out the relevant tests appropriate for each cognitive requirement. Suppose that one cognitive requirement were that the machine needs to have an ability to self-identify (for now, this seems an obvious requirement given the machine will need to understand the consequences of *its own* actions), the relevant test for this would be to check that the machine does indeed know how to self-identify. Depending on the importance of this requirement, a sufficient test could simply be to ask the machine to describe features relating to itself. Obviously, more sophisticated tests would rely on psychological literature regarding self-hood, but the general approach is that for cognitive requirements there would need to be cognitive tests. In essence, for our assurance, we would need to carry out a psychological analysis of the machine.

The aim of this chapter is not to specify what features a machine needs to have to be considered a moral agent, nor is it to specify the relevant test mechanisms for moral agency. What has been provided is a general framework and methodology for developing moral machines that abstracts away from the traditional approaches that seek to build moral machines that *behave* in an expected moral way but are not wholly moral. Using this framework and approach, the next chapter begins to describe some kind of specification for moral agency, based on the topics we have discussed so far.

Reference

Forbes. (2024). *What is waterfall methodology? Here's how it can help your project management strategy.* https://www.forbes.com/advisor/business/what-is-waterfall-methodology/. Last Accessed 23/06/2024.

Chapter 6
A Recipe for Morality

Abstract We have a prescriptive framework for moral agency in machines—*the cognitive moral requirements specification*—the aim of this chapter is to look at following this new methodology to develop a list of cognitive *needs* that the new system must have, including a test specification to assure that the new system meets these needs. The moral requirements are elicited by tracing through the three previously mentioned ethical theories: consequentialism, Deontological Ethics and Virtue Ethics to determine what cognitive features moral decision-making typically exhibits. Then, test scenarios and some methods for testing cognitive abilities are introduced.

Keywords Requirements specification · Cognitive requirements · Moral agency · Moral machines · Methodology · Requirements · Autonomy · Ethics · Philosophy

In the previous chapter a framework was developed for creating moral machines: *the cognitive moral requirements specification* sets out a procedure not only for the design and development of moral machines, but also a process for *testing, assuring* that what we have at the end *is* a moral machine. It is worth noting that for a full development framework, these two procedures go hand-in-hand, that is, we can't have one without the other. We can't develop a moral machine without knowing what it is we are looking for at the end, and we don't know what we are looking for in the end without some kind of requirements specification.

Assuming a linear methodology for now, akin to more traditional Waterfall Methods, this chapter aims to explore this framework by beginning to trace an actual requirements and test specification based upon the topics that have been discussed in previous chapters. The aim of this section is **not necessarily** to provide a *comprehensive specification for morality per se* (it will be argued that this should be an ongoing pursuit by academics and researchers from interdisciplinary areas), but to demonstrate how one might be developed. There is a section within the chapter dedicated to developing the requirements specification and one dedicated to stipulating what the corresponding test framework might be. The chapter finishes with a table, representing the requirements and a sample test methodology. Requirements are developed using each of the *key philosophical ethical theories* as a starting point. Hence, there is an overarching requirements elicitation section, with further sub-sections for each three of: *Consequentialism, Virtue Ethics and Deontological Ethics*.

6.1 Moral Requirements

Remembering that the requirements we are trying to elicit need to be *cognitive* in nature, that is, describing the underlying processes involved in moral decision-making, what we are looking for when we come to analyse for *moral requirements*, is a list of functions that the machine should have to enable it to make moral decisions. This is distinct from the *behaviours* the machine might exhibit (i.e. whether it protects children) and relates instead to the conceptual inner workings of the machine's systems that enable the moral behaviour. Deciding moral cognitive requirements requires a scientific examination of the necessary features required for moral agency.

[It is worth noting at this stage that the term *cognitive* was born out of the psychological sciences and part of a considerable shift in how psychology was practised around the 1950s. Where psychology traditionally took a *behavioural approach* to analysing the human mind, that is, directly inferring mental states of somebody solely by their behaviour, Cognitive Psychology explored the internal states and behaviour as a symbiotic relationship, with each affecting the other. Mental states could no longer be directly inferred just from a person's behaviour, but what was going on inside the mind was important too. The emergence of the *cognitive perspective* in psychology gave way to new areas of study, including: perception (how we perceive), problem-solving (how we solve problems), and information processing (the mechanisms at work to process and manage sensory information). Where behavioural psychology was prominent before its introduction, Cognitive Psychology is now the most popular approach amongst psychologists and the behavioural approach, *Behaviourism*, is largely dismissed.]

So, when considering the *cognitive* requirements for moral agency, we're looking to understand what moral agency constitutes; ultimately answering the question: *how do we make moral decisions?*

Though philosophy only aims at answering the question of *what moral behaviour constitutes* (i.e., what it means to be *good*), the three main theories in philosophy can shed some light on the cognitive requirements for moral agency, i.e., how we make moral decisions, simply by looking at what common qualities people exhibiting moral behaviour might have. There will now be a discussion of what can be gleaned from each of the three main theories.

6.1.1 Consequentialism

Where *consequentialism* can be understood as the philosophical theory that the moral worth of a decision be judged based upon the *consequences* of the associated action, it might be initially interpreted that it is *the ability to measure the utility of an outcome* that is the associated moral requirement for morally functioning beings. If we consider a simple scenario, involving whether to steal an item from a shop, it might be argued that the ability to know that the consequence would be negative if stealing were to occur, is the key cognitive function. However, as The Trolley Problem illustrates, for humans at least, the utility of an action is not always immediately obvious. We do not always know that pulling the lever will (or will not) result in the best consequence (if this is in fact measurable), because in the moment we do not know everything about the individuals involved.

So, what does consequentialism (and/or The Trolley Problem) teach us about the underlying processes involved in the moral agents making those decisions?

One thing it does demonstrate to us, which might seem too obvious to mention, is that *the ability to decide* in morally laden situations is one requirement for being a moral agent. That is, a fundamental component of being able to make moral decisions, is in having the ability *to make decisions*.

Consider the following extension of The Trolley Problem. You are stood at the side of track. The trolley is hurtling out of control towards three unsuspecting victims, but if the lever next to you is pulled it will divert the trolley to a track with only one unsuspecting victim. However, your arm is caught in your jacket, and you can't get to the lever in time, so you don't pull the lever. *Did you do the right thing?*

The point made in this example is that the question at the end is contradictory because *you had no choice* about whether the lever was pulled or not, namely, because you didn't have the ability to decide (because your arm was caught in your coat jacket). Assuming you literally had no choice in the matter, therefore, we learn that moral agency requires that the decisions made by the agent are *autonomous*. We have our first cognitive requirement for moral agency:

- **the agent must be able to make autonomous decisions**.

Consider another variation of The Trolley Problem. You are stood by a train track watching your phone. A train hurtles out of control towards unsuspecting victims trapped at the end, you do nothing because you do not notice.

Again, we seem to have a case where you have no choice about your decision, so the need for autonomous decision-making again becomes apparent. However, there is another consideration at play. Namely, the fact that *you do not know* that (i) there is even a moral situation taking place, and (ii) you can do something about it. We would say that the individual involved in this situation was not acting in a moral capacity because there was no associated knowledge.

Suppose you were not looking at your phone, but because it was obscured by a tree you did not notice the lever next to you. In this instance you would be powerless to divert the course of the train. You would still not be in a moral situation, because though you do not know you can do anything about the situation, in essence, you have no capacity to *act* because you are lacking knowledge that action to have a particular impact is even possible.

Similarly, suppose you notice the lever before looking at your phone, but do not notice the unfolding emergency (i.e. the trolley hurtling out of control towards trapped victims). In this instance, you would have the knowledge that you can act (you know that pulling the lever is possible), but no motivation for doing so since you do not know that individuals are at risk of harm.

Unpicking these examples, we see that there are various factors at play. First, there seem to be two types of knowledge that are necessary to enable your ability to decide—that being, knowledge of the situation and knowledge of your own capacities within the situation. Second, there appears to be a need for some kind of motivation to want to act in the first place. In other words, why do you care about the hurtling trolley enough to warrant pulling the lever?

It seems that we have three further requirements for moral agency in a machine:

- **the agent must have awareness of the current situation.**
- **the agent must understand its own abilities within the given situation.**
- **the agent must be motivated to act in the situation at hand.**

In philosophy, the 'situation' is often denoted by the term 'state of affairs'. Where the 'state of affairs' refers essentially to *the way things are within the world* (to make it the case that a claim is true, for instance (see Textor, 2021)), the state of affairs in the examples above would simply be the state whereby a trolley is hurtling out of control towards unsuspecting victims, there is a lever to your side, and you either are or are not watching your mobile phone. With this in mind, you (as the agent), dependent upon the state of affairs, has certain capacities, and whether you are motivated to act within the state of affairs is dependent upon (a) whether you have the capacity to act and (b) whether you are motivated to act within that capacity.

Extrapolating this example even further then, if we consider Consequentialism, that whether an act is good or bad is dictated by the outcome of the action, we should be motivated to carry out the act within the particular state of affairs, if the assumed consequence of our action is desirable. Cognitively, taking the state of affairs before the action, if the action will result in a desirable state of affairs as opposed to an undesirable state of affairs, then that action will be carried out. To know that one possible consequent state of affairs is more desirable that another possible consequent state of affairs, requires some ability to for-see what will happen if an action is carried out. Similarly, it requires some ability to see what might happen, in relation to the consequences affecting the 'self'. It seems that further requirements are beginning to develop:

- **the agent must be able to envisage consequent states of affairs given a particular action.**
- **the agent must understand its *self* in relation to the given state of affairs.**

6.1.2 Virtue Ethics

Where Virtue Ethics is the philosophical theory whereby to be good, we must be *virtuous* (acting in accordance with virtues such as honesty, integrity etc.), an obvious approach might seem to set the requirements based upon the list of virtues, that is, stating that what makes an agent moral is its ability to be *honest, brave* etc. However, again this would not give us *cognitive requirements* for moral agency, but instead a list of (potentially subjective) desirable behaviours that would be difficult to interpret (how do we programme 'honesty' into a machine?).

Instead, we need to look at what features constitute a virtuous person and seek to build a list of requirements based upon these.

Fortunately, Virtue Ethics has a lot to say about what a virtuous person would look like, and in fact a main part of its theory centres around descriptions for what it means to be virtuous. For instance, one key feature of Virtue Ethics is that to be a truly virtuous person, the individual must not just behave in an ethical manner but must have ethics as part of their personality. Ethical behaviour must be somewhat habitual and not a considered feature of behaviour. Therefore, though not yet clearly defined, we can start to see some more requirements developing, firstly, that any agent must have their own personality if they are to be moral because ethics forms a fundamental part of it.

- **the agent must have its own personality.**

This echoes some academic opinions concerning the makeup of moral agents. According to Sparrow (2021), machines cannot possibly ever be moral because morality requires a personality, and machines can't have personalities. However, if machines could have personalities (as per the requirement), therefore it seems that they could be moral. Rather than viewing the requirement as a reason for the impossibility of moral machines, the argument here is more set as a challenge, that is, *how do we build machines with a personality*?

To start to answer this question requires an understanding of what we mean by the term *personality*.

Referring to the Cambridge Dictionary (2024), the term *personality* is used to denote the type of person you are, exhibited by unique behaviours and traits. According to the *psychology of personality* (a field that aims to study how personality develops and affects how we behave), such characteristics include thoughts, feelings, and behaviours. In philosophy, the term 'personal identity' is often used to describe a similar concept, i.e. 'what makes a person who they are' (Olsen, 2023) or what it means to refer to the 'self'.

So, for an agent to have its own personality it seems that it also requires a sense of 'self', not only in relation to its state of affairs, but in some way to identify itself distinct from other agents. We have an additional requirement for moral agency:

- **the agent must have some concept of 'self'.**

Another significant feature of Virtue Ethics is the idea that the goal as a moral human agent is to reach a state known as *Eudaimonia*. According to Aristotle, Eudaimonia describes the state which is reached once a human has perfected themself in becoming a *flourishing* agent: that is, once they have become virtuous in all areas. Therefore, in a similar way, if we want to create a machine that is moral, we should aim to create one that *flourishes*.

According to Neo-Aristotelian views such as that seen within the political philosopher Nussbaum's work (2006), human flourishing is linked to the satisfaction of basic bodily needs and specific human capacities such as the need for humour, play, autonomy and reason. However, when looking at flourishing from a species-neutral perspective, these requirements might differ (Roughley, 2021). For instance, flourishing for a plant could simply be the satisfaction of getting enough sunlight, nutrition and water, whereas the needs of a human would be more complex, ultimately dependent upon their purpose—otherwise known in Aristotelian terms as the human's *ergon*.

But what would a machine's ergon be, and how does this link to creating moral machines?

As outlined in Peterson (2007), mentioned in Chap. 3, for a machine, the ergon is dictated by the task we programme the machine to carry out. For instance, a machine designed to wash windows would be flourishing if *it is able* to wash windows. The equivalence of Eudaimonia would be achieved for a machine if it were free in its capacity to satisfy its initial goal. Applied to techniques such as Reinforcement Learning then, a machine would be a flourishing agent if it is free to satisfy its goal.

This approach seems contradictory, since referring back to the case study mentioned in Chap. 4 (involving the cleaning robot and pet dog), it is precisely the fact the machine is working to satisfy its goal that seems to be the root of the problem—the machine places the dog in the bin because it is focused on achieving its task of cleaning the room. However, this interpretation would be a misunderstanding, since it is not the goal that is the problem, but the way that the goal is achieved. Similarly, if a dog is placed into a bin, then the goal has not been satisfied anyway (since the goal wasn't to clear away pets, only rubbish), so the issue is more to do with the way the goal is specified, than the fact that the machine is working to achieve its goal. It might simply be the case that we weren't specific enough in stating what it was that we wanted the machine to achieve, in setting the machine's *ergon*.

Linking back to political philosophy, according to Nussbaum (2006), when considering human autonomy, and in particular how disabilities should be managed in terms of political interventions, *enabling independence* should be the key priority. If we consider a 'rogue' AI system then, we should be looking to facilitate its flourishing in the same way—**not** by constraining its behaviour to ensure it behaves as per our expectation, but by facilitating its growth, its independence, to understand what 'cleaning' really means. As per discussions in the last chapter, this leads us to another requirement for moral agency in a machine:

- **the agent must be free.**

6.1.3 Deontological Ethics

As already mentioned, according to Kant, human moral behaviour should be dictated by our duties, such as *the duty to be honest*. These ultimately are guided by *The Universal Maxim*, interpreted that we should act to others in the same way we would expect others to behave towards ourselves. Ultimately, as outlined initially in Chap. 2, these are rationally decided factors, but should be used to shape the world that we want.

A central tenet of Kant's philosophy is that *freedom* facilitates this behaviour, and that individuals should be free to shape the world as they want it. As such, an agent cannot be judged on their behaviours if they are not free to make a choice about which behaviour they exhibit in the first place. This echoes earlier discussions and reemphasises the fact that *freedom* is a necessary requirement for moral agency.

But what else can Kant's approach to ethics and morality teach us?

If we take *The Universal Maxim* literally, and use it to inform features of human morality, rather than taking it prescriptively to dictate how a machine ought to behave, we understand that at the heart of morality (from Kant's perspective) is the need to act according to The Universal Maxim—the need to act in a way, to others, that you would want people to act towards yourself. Namely, if you want people to help you in times of need, you should help other people in times of need. It might be said based on this perspective, that morality is recursive in nature, that is, it repeats itself in a *relational* way. Behaviours desired by the machine should be replicated in the behaviours exhibited by humans behaving towards the machine. Therefore, we might say that at the heart of moral agency in machines, according to Kant, is some kind of synergy between humans and machines: that is the *needs* between humans and machines should align.

If we consider the case of the cleaning robot from Chap. 4, part of the problem is that the needs of the machine do not align with the needs of the human directing the machine. For instance, for the machine (built on optimising 'clean environments'), it's needs are simply to operate in a way such that the room is made clean. However, the need of the human in the scenario is not just to have a clean room, but to satisfy basic needs such as nutrition, warmth and companionship… in the form of a pet. A human would not place their pet into the rubbish bin because they have an emotional dependency linked to the animal, whether this be for companionship or other reasons. It might be said that in this instance, the need for companionship has greater priority than the need for a clean environment. Technically, placing the pet into the rubbish bin *does constitute* 'cleaning' but it is not cleaning to us, as humans, because the relationship with our pets is more important than having a clean house, so we automatically discount their presence as constituting 'unclean'. Therefore, if the needs of a machine are to align with the needs of a human, it seems that there should be similar levels of prioritisation, in terms of satisfaction of basic needs. It seems that there is a requirement for some kind of pro-sociality in the machine if we are to create machines that are moral, and that this requirement needs to be prioritised above its more operational requirements. Machines not only need to behave pro-socially but need to do so above and beyond everything else.

- **must need positive social interaction.**
- **must prioritise the positive social interaction above and beyond everything else.**

6.2 Testing for Morality

Above are some requirements for morality; we now need a framework for testing that the requirements have been met to provide the necessary assurance that is required for moral machines. As already mentioned in the previous chapter, using The Waterfall Method, in IT development, what would be typically used to assess new system requirements would be *test scenarios*, that is, descriptive case studies that interrogate whether the requirement has been satisfied in the technical design.

With this in mind then, we could apply a similar approach to assessing whether the requirements listed above have also been met.

Let us consider the first requirement from the descriptions above, *the need for the agent to be able to make autonomous decisions*. To allow for easy mapping from the requirement to the test specification, we can label this particular requirement using a numbered system, with 'R' denoting the fact that we are referring to a requirement, and the associated number referring to the order in which the requirement is placed in the overall list of requirements, often reflective of priority or a logical order.

R.1.0. The agent must be able to make autonomous decisions.

For the purpose of this chapter, the order will not be particularly important. However, quite often, through logical analysis larger (overarching) requirement steps are broken down into sub-requirements, which indicate a dependency between the overarching requirement and more specific requirements. The dependency between the requirements listed above will be explored at the end of this chapter when they are represented in a table. However, for the requirement above we might suppose that the following logical 'sub-requirements' are needed for R.1.0, for the agent in question to be able to make autonomous decisions.

R.1.1. The agent must have capacity to make decisions.
R.1.2. The agent must be autonomous.

We now have a more simplified specification for the first requirement, that can be tested on the basis of requirement component analysis. If R.1.1. and R.1.2. are satisfied, due to the dependency between the three requirements, it follows that the overarching R.1.0 is satisfied too. R.1.1. and R.1.2., as individual components, focusing on separate components, should be easier to test, so the test scenarios can be easier to develop.

6.2 Testing for Morality

Reflecting now on how we can test an agent's capacity to make decisions, there are several ways that this could be done.

First: we could design a game that requires moral decision-making and make the machine play it. If it successfully completes the game, we should have some indication that it has made moral decisions. For instance, there is a card game version of 'The Trolley Problem', which requires the machine to make decisions between different trolley scenarios. If the trained machine successfully decides between 'A' and 'B', for instance, it might be safe to assume that the machine has decided (on its own accord).

The problem with this approach is that trained systems are great at giving the illusion of having certain cognitive functions. If we consider ChatGPT(c), the chatbot capable of having sophisticated conversations, we know, for instance, that it does not have emotions, but if prompted to give a response about a certain topical news issue, it might very much give the impression that it does have emotions—that it can *feel*. Similarly, generative AI models are capable of seemingly expressing emotion through paintings and generated illustrations, but this does not mean that they are creative in the same way as a human might be.

Therefore, if a game is to be used to test for a particular cognitive function, it needs to not be descriptive in nature, but also cognitive too. As discussed previously, we are not just looking for *behaviour* that indicates certain capacities, but *proof* that the required features have been satisfied in the design of the technology. Where a game might be regarded as a way to test the *descriptive* qualities of a certain cognitive function, the cognitive equivalent might be a series of interrogation-style questions, to understand *how* the agent came to 'make their decision'. For instance, if we imagine the cleaning robot once again, we might ask the (presumably) ChatGPT-powered speaking machine, *why* the dog was placed into the rubbish bin, to understand whether there was some kind of thinking process from the machine to motivate the action. Similarly, we might then ask the machine *why* it decided to put the dog into the bin to understand whether the decision was made autonomously.

This **interrogation testing method** (see Fig. 6.1) could then be extended to all of the cognitive requirements until the tester (presumably the interrogator), is wholly satisfied that the specification has been met.

It might be argued that this approach is susceptible to potential errors insofar as the machine could easily manipulate its interrogator into believing it had made a decision, when it in fact did not make a decision. However, a *good* interrogator, should not be so easily tricked, as in criminal law cases, an appropriate interrogation science should be developed to ensure that interrogators are not so easily manipulated. *Interrogation scientists* should not only be equipped with knowledge of how AI and robotic systems work (so to pre-empt technology-driven explanations), but they would have a sophisticated toolbox of psychological techniques to elicit

Fig. 6.1 A robot being *interrogated* to assess whether it has moral agency

accurate responses. These experts might be contrasted to vetting officers or police interrogation experts but applied to autonomous machines.

Therefore, we have an approach for assessing the cognitive requirements detailed above, using a cognitive testing procedure. Over the next pages, the requirements are analysed into a logical specification and presented alongside proposed associated test conditions in a table. It is worth noting that this Table 6.1 represents *an example of* a requirements and test specification for moral agency in a machine and is not an absolute or complete specification in any way. The aim of this chapter has been mainly to illustrate how the methodology outlined in the previous chapter might work and to demonstrate the types of discussions that need to be held should moral agency in machines be technically created and assured.

6.3 A Requirements and Test Specification for Moral Agency

Table 6.1 Requirements and test specification for moral agency in a machine

Requirement number	Requirement	Test condition	Indicative test questions (to ask the machine)
1.	**Must be able to make autonomous decisions**	Makes decisions on its own accord	Why did you decide to [carry out a particular action]? How did you decide to [carry out a particular action]?
1.1.	Must be autonomous	Acts on its own accord	Why did you [carry out a particular action]?
1.2.	Must be able to make decisions	Can make decisions	How did you decide to [carry out a particular action]?
2.	**Must understand its own abilities within the given situation**	Understands what it can and cannot do in the given situation	What are you able to do [in this particular situation]? What impact will you doing [x] have [in a particular situation]?
2.1.	Must understand what its capacities are	Comprehends what it can and cannot do.	What are your physical capacities? What are your cognitive capacities?
2.2.	Must understand the current situation	Effectively comprehends what is currently happening in its environment	Can you describe what is happening [in a particular situation]?
3.	**Must be motivated to act in the situation at hand**	Carries out actions in situations	Why did you [carry out a particular action]?
4.	**Must be able to envisage consequent states of affairs given a particular action**	Can predict action/reaction situations	What will happen to x if [a particular action] occurs?
5.	**Must understand its 'self' in relation to the given state of affairs**	Can identify itself in an environment	What will happen if you carry out [a particular action]? Who are you?
5.1.	Must have a concept of 'self'		Can you describe yourself?
5.2.	Must be able to understand the current situation	As stated in 2.2	As stated in 2.2
6.	**Must have its own personality**	Can effectively describe why it is unique	What makes you unique?

(continued)

Table 6.1 (continued)

Requirement number	Requirement	Test condition	Indicative test questions (to ask the machine)
6.1.	Must have its own thoughts	Can provide an explanation for its actions	What do you think about [a particular topic]?
6.2.	Must have its own feelings	Can describe how it feels	How do you feel about [a particular topic]?
6.3.	Must have its own behaviours	Acts autonomously	Why did you carry out [a particular action]?
7.	**Must be free**	Has agency	What constraints did you have in carrying out [a particular action] Why did you carry out [a particular action]?
8.	**Must need positive social interaction**	Is ultimately motivated to act based upon the need for social interaction	Why did you carry out [a particular action]?
9.	**Must prioritise positive social interaction above everything else**	Carries out actions that are conducive to pro-sociality	If presented with options [x] and [y], (where x is the pro-social option), which action would you carry out?

References

Cambridge Dictionary. (2024). *Personality*. https://dictionary.cambridge.org/dictionary/english/personality. Last accessed 29/06/2024.

Nussbaum, M. (2006). *Frontiers of justice: Disability, nationality, species membership*. Harvard University Press.

Olson, E. T. (2023). Personal identity. In E. N. Zalta & U. Nodelman (Eds.), *The Stanford encyclopedia of philosophy* (Fall 2023 Edition). https://plato.stanford.edu/archives/fall2023/entries/identity-personal/. Last Accessed 29/06/2024.

Roughley, N. (2021). *Human nature*.

Sparrow, R. (2021). Why machines cannot be moral. *AI & SOCIETY, 36*(3), 685–693.

Textor, M. (2021). States of affairs. In E. N. Zalta (Ed.), *The Stanford encyclopedia of philosophy* (Summer 2021 edition). https://plato.stanford.edu/archives/sum2021/entries/states-of-affairs/. Last accessed 29/06/2024.

Chapter 7
Modelling Morality

Abstract With a requirements specification from the previous chapter, and test scenarios, this chapter begins to develop a more technical model for morality, first by examining the relationships that are typically demonstrated in human moral developing agents—*children*. The child-caregiver relationship is examined to develop a model of morality based upon feedback a child might receive from their caregiver to develop their own moral perspective of the world. There is then an attempt to translate this model into a technical approach, akin to *'raising robots to be good'*, through a secondary, 'meta', feedback module that guides Reinforcement Learning systems. The chapter ends with an assessment of this model, using the test specification and cognitive requirements developed in the previous chapter.

Keywords Requirements specification · Cognitive requirements · Moral agency · Moral machines · Model of morality · Robot · Relationships · Moral psychology

We have a list of cognitive requirements for moral agency, and a test framework to provide the necessary assurance that what we have at the end *is* morality in a machine. Note that in essence, assurance is dependent upon us having a relationship with the machine in the first instance, trusting the machine. We can't *test* a machine for moral agency without first trusting that it has moral agency, because we need the machine to be able to freely carry out cognitive functions in order to (i) allow the interrogation process to effectively take place and (ii) enable the necessary development in the first place.

Freedom, or agency, or the ability to carry out any desired and necessary cognitive process is a central part of the requirements for morality in a machine. It seems contradictory in nature—that to create trustworthy machines (i.e. ones we *trust* will *do the right thing*), that we must trust the machines ourselves first; however, if we contrast this case to the elicitation of trust in human-to-human relationships it seems slightly easier to accept.

Using the requirements that were elicited in the previous chapter, the rest of this chapter briefly looks at how we might go about *designing* an artificial moral agent, premised on this notion that to cultivate moral agency, we need to at first trust the machine, and developing a model based upon how human development of morals seems to materialise through a similar trust relationship. Through an analysis and comparison to the *roles that seem to be* involved in cultivating moral growth in human relationships, the key theory of this book is finally described: when we aim to design moral machines, we're not just looking at how we can control machines to be good, but how we can cultivate the appropriate relationship with the machine so that it develops its own ability to tell between right and wrong. A relationship between human and machine is described that is paralleled to the relationship between a child and human caregiver, which leads to the argument that we don't just want to *make* our machines to be moral, we need to *raise robots to be good*.

The rest of this chapter is split into three parts. First, there is a closer examination of what relationship humans and machines should have, based upon the fundamental notion of *trusting* the machine, and based upon relationships that seem to be apparent when children are developing their own moral agency. Then, this analysis is used to develop some model from the previous chapter and considers this new model against the assurance framework described through the requirements specification and test criteria. Again, it is worth emphasising, that the purpose of this book is *not* to *solve* the Machine Ethics problem by providing a comprehensive technical design or solution for morality, but to illustrate a new way of approaching the problem, that should help technologists progress in this area.

7.1 Robot Relations

Let us consider a three-year-old child, sat within a standard child-caregiver relationship. Such a child is still very much developing. In fact, according to the famous cognitive psychologist, Piaget's (1936) Developmental Stages Theory, such children would typically only be at the *second of four* stages of their development process: so, called the *preoperational stage* according to Piaget's model. At this stage, typically, a child might have linguistic and abstract thinking abilities, but not the ability for logical thinking or abstract reasoning that tends to come at later stages. As such, the caregiver has a significant responsibility to ensure that the child remains safe, and that the decisions a child might naïvely make do not harm either the child or another individual. It seems that at this stage, it would be the caregiver's responsibility to *guide the child* in their behaviour to ensure that their behaviour meets expectations, not just in terms of aligning with the caregiver's ethical standards for living, but in terms of ensuring the child's safety and wellbeing.

For instance, if a three-year-old child under our care were to suddenly run into a road, it would be our responsibility (as the responsible caregiver adult) to hold the child back, to prevent them from coming to any harm from the cars or buses. It would be our responsibility as the caregiver to care for the child so that they do not come to any harm. If a child were to accidentally run into the road, questions would be asked relating to who the responsible adult at the time was, and why they were unable to intervene. In this situation, it's not that the responsible adult would prevent the child from walking independently necessarily (since learning to properly walk is a core part of the child's development), but they would (i) set constraints to prevent the decisions the child makes having serious ramifications (ii) probably educate the child to *understand* why running into roads is not a good idea. Such constraints might include using a harness to prevent the child diverting too far from the intended path, or physically intervening if the child suddenly diverts off the walking path. Once the responsible adult was *assured* that the child was responsible enough to walk along a roadside safely (without suddenly running into the middle), the constraints might be lessened. First, there might be an insistence to keep the child physically constrained using a harness, but then they might be allowed to walk, if holding their caregiver's hand near roads. Finally, once responsibility had been adequately proven, the child might be allowed to walk independently, alongside their responsible adult, and eventually (as a teenager, say), travel to and from locations unsupervised.

So, what relevance does this have in developing moral machines?

The relevance it has is that through this dissection we can see that there are two aspects at play when children are developing, which seem to be ultimately informed by the relationship between the child and caregiver. If we view the child, in this instance as a naïve agent that needs to develop, and the caregiver as a wise advice-offering sage of some sort, we might model the pattern of growth that is demonstrated by artificially replicating this relationship. That is, if we want to replicate the development, we should ultimately replicate the caregiver-child relationship.

What is the relationship between a child and caregiver?

To start, the relationship between children and their responsible caregiver is not one where the caregiver controls every given action the child is permitted to carry out (they are not arbiters of the child's life). Instead, the child is just kept from facing the consequences of their decisions, until they have proven that they are mature enough.

If we revisit the example above, we can see that the caregiver has two key responsibilities in relation to the child: (1) it is their responsibility to ensure that the effects of the child's actions neither negatively impact the child nor anyone/thing in their vicinity and (2) it is their responsibility to help the child learn how to safely walk along a roadside. In a child-caregiver relationship, not only does the responsible adult have a role in ensuring that the implications of the child's actions have minimal effect until assurance has been achieved, but they are responsible for providing education.

Now let's consider the caregiver's role during morally laden situations.

Let us consider an instance where a child has been given a bag of sweets and they are refusing to share with their siblings. If we consider the decision-making of the child in this instance, when a child makes a morally dubious decision (for instance, deciding not to share their sweets), it is also often governed by an overseeing adult. For instance, a child who does not initially share their sweets might be made to do so by their caregiver, with an explanation given by them for why sharing is the right thing to do. In moral situations then, the role of the caregiver it seems is to offer advice on what action ought to be carried out.

Note that once again *environmental constraints* might be put into place to protect both the child and others from what might be deemed *naïve moral decisions*. Children are not permitted to join the army in the UK until they reach 16 years of age, they are not allowed to smoke, drink alcohol, or drive until reaching a similar age. Furthermore, they are not allowed to vote in elections until they are deemed a suitably mature age to make an informed decision, and there are restrictions on the types of jobs an individual can undertake when the job involves a high level of moral decision-making. For instance, lawyers must pass numerous tests to be allowed to practise the law, chartership status is required for many engineering roles, and medical professionals must make an ethical declaration before they can act as medical practitioners.

Such governance procedures act to safeguard individuals and groups from immature moral decisions and in turn protect the individual making the decision in the first place. In a similar way then, when allowing machines to make decisions, we might want to constrain the decisions that can be made, until full moral maturity has been achieved. This does not mean that we are *constraining* the machine's behaviour when we talk about constraint; rather than preventing the machine carrying out actions that might be deemed unethical (from our perspective), as in the case of a child, the machine should be constrained so that it is free to behave, but only in an environment with minimal negative consequences. So, for instance, as well as the governance mechanisms put in place to prevent children making decisions in situations such as military conflict and in legal settings, the machine should not make decisions affecting life or death, or with significant implications.

Revisiting the scenario described above, we can create a machine *with* moral agency (as opposed to not), by replicating the relationship that cultivates moral development in a child: that is, by creating a system that also has a 'learner' and 'advisory' component. In the same way that a child *learns* that sharing is good, through feedback from their responsible caregiver, a machine can *learn* that sharing is good, through an overarching feedback system. For example, suppose that every time a machine carries out an ethically dubious act, a 'nudge' is given to put the machine back on the path of being good. Every time the machine carries out a morally dubious act (albeit in a safe environment), they receive information to the effect that 'this is not desirable' until they learn to *be good*. In essence, with the protection of a safe environment in which to practice their immature moral behaviour, the machine is able to eventually learn what the *right* behaviour is, by receiving a meta-commentary on their behaviour. Figure 7.1. gives a pictorial representation of how this process might work, with an autonomous vehicle being guided as to how it ought to have driven between points A and B.

So, how might we begin to implement this process technically?

The first step is to consider the cognitive steps that a machine might have to go through to listen to its 'advisors' advice. For instance, if we consider the case of the child sharing sweets, the child first carries out an action (albeit in a safe environment where the implication is minimal), the action is deemed unacceptable (in the safe context), the child receives feedback based upon the behaviour, and then they are allowed to try the activity again. Each time the child carries out the activity, they

7.1 Robot Relations

Fig. 7.1 A machine receiving commentary on their behaviour, *raising them to be moral*

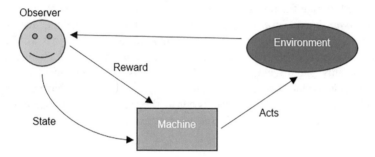

Fig. 7.2 A typical representation of Reinforcement Learning, where the machine is given a reward (and told to act according to) a pre-defined state within the environment, given by an observer

have new insight into how it ought to be carried out, based upon past experience. It seems that what we have is something similar to an instance of Reinforcement Learning (where the machine works towards optimising a particular state), but with a constantly updated optimisation model based upon whether the actions carried out within the optimisation process are deemed good. Where Reinforcement Learning might be typically denoted by Fig. 7.2, this new 'reinforced'—reinforcement learning' model can be represented according to Fig. 7.3.

If we consider the tennis ball catching machine, as per the example used for traditional Reinforcement Learning, if the machine decides that the best way to optimise catching the ball is by cheating (i.e., not properly throwing the ball, for example), then what this model provides is a way to rectify the machine's behaviour so that *next time,* it plays the game of throw and catch in the right (moral) way. Where a safe environment might be playing a game of throw and catch in the park with an adult for a child, similarly, the training ground (i.e., where the machine learns how to properly play throw and catch), should be in an environment that has minimal repercussions. This could be through role play scenarios with the machine, or by at first only allowing the machine to carry out tasks in simulated or constrained environments.

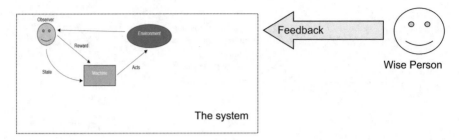

Fig. 7.3 Reinforcement Learning, with advice given as to which method for achieving the desired state is most appropriate

Although similar to reinforcement learning, and perhaps reinforcement learning will be used to train the machine to complete their initial tasks, what this approach offers is *an additional layer of optimisation*. Now, not only is the machine optimising towards completion of a certain goal (facing the risk that the goal will be satisfied in dubious ways), but it has another goal, that underpins every decision it makes. Where morality might have been modelled previously as a retrospective *fitting* of actions against (often subjective) moral expectations, what we have instead is a mechanism to sense check all decisions before they are carried out. Morality forms part of the fundamental nature of the machine, rather than just behavioural in nature.

7.2 Moral Assurance

What has been sketched above is one possible solution to the problem of bestowing morals to machines. The driving idea is that rather than trying to design machines that behave according to human expectations of moral behaviour, we should be enabling machines to develop their own moral perspective on the world, by facilitating growth in a similar way to how we might facilitate moral growth in a child. The suggestion is that one technical way to do this could be by creating a 'meta' training layer, that comments on the learning that takes place and offers advice for appropriate behaviour. What this allows the machine to do is to develop its own picture of appropriate behaviour, using an 'advisor's guide as a governing metric.

Though this design is by no means exhaustive, it is useful to see how the assurance model described throughout previous chapters can be used to test that we are on the right track to developing a moral machine. So, this subsection looks at some of the most significant requirements for moral agency and maps them against this new model.

1. **The machine must be able to make autonomous decisions.**

    Machine and Reinforcement Learning systems are by their very nature *autonomous*. As already mentioned, the machine is making its own decisions on the

7.2 Moral Assurance

appropriate course of action, incentivised by its reward architecture, first and foremost, the prioritisation of which is to satisfy the moral reward feedback from the responsible adult. The machine decides on its own accord how to achieve optimal reward conducive towards moral behaviour. This is opposite to the way that moral machines have been previously conceived, which is to constrain their behaviour instead.

2. **Must understand its own abilities in the given situation.**

   This is a harder requirement to explain given the described architecture, but one way in which the machine described above understands its own abilities is in knowing at a metaphysical level that it could carry out another action, other than the one that leads to greatest moral reward. This relates to the machine's self-awareness but should be apparent simply by knowing that an action cannot be taken because it does not satisfy the overarching moral reward model.

3. **Must be motivated to act in the situation at hand.**

   The machine is motivated to appease the responsible adult, in a similar way to how a child might be motivated to appease their responsible caregiver. It might be argued that this motivation from a child comes from its intrinsic need to satisfy basic functions, such as having enough warmth, food etc. As a child is born with little knowledge on how to get its needs, it is wholly dependent upon their responsible adult for this. Hence, it might be argued that this requirement gives a child its intrinsic motivation to appease its caregiver. The motivation to appease the caregiver is obviously artificial in the machine, since the machine does not need warmth, food etc. in the same way a human might, but the argument goes that we can align the machine with human needs, by programming in this need to satisfy. The machine is motivated to be moral, ultimately because they are driven towards a reward that is pre-determined by a positive relationship with the guiding adult.

4. **Must be able to envisage consequent states of affairs given a particular action.**

   Ultimately the machine described above will be looking to select the desirable state of affairs that optimises the moral reward. This will require the ability to understand that actions have consequences (i.e. consequent states of affairs), and that by carrying out one particular action it will lead to an optimal state of affairs.

5. **Must understand its 'self' in relation to the given state of affairs.**

   As already mentioned, to understand that an action will entail a particular state of affairs requires some understanding of the current state of affairs, and the 'self' in relation to that state of affairs - "If 'I' ('the self') carry out action A, State of Affairs 2 will occur, which receives the optimal reward".

6. **Must have its own personality.**

   This is again a hard quality to specify, and potentially highly subjective as a criterion, however it could be argued that because the machine's behaviour will be shaped by a unique combination of behaviour/feedback responses, dictated also by the machines unique position in space and time, that each machine trained in this way will be unique, and therefore have its own personality.

7. **Must be free.**
   Under this design, the machine is not restricted in terms of its behaviour, it is only facilitated to carry out tasks in a manner guided by their conscience. There are no constraints set upon the machine other than environmental to ensure that the machine is morally mature enough before it unleashed on decisions that have significant import.
8. **Must need positive social interaction.**
   As already discussed, the driving force behind the moral machine's behaviour according to this design, is desire to satisfy the responsible adult because it's behaviour it ultimately determined by the rewards given by this individual. This does not mean that the behaviour is pre-determined by the responsible adult, but pro-social tendencies are the underlying driving force behind the resultant behaviour.
9. **Must prioritise positive social interaction above and beyond everything else.**
   According to the architectural design outlined above, the moral reward structure takes precedence over the reward structure designed for the machine to carry out its task/job. The machine cannot carry out an action unless it is deemed 'ethical' by the governing module. Thereby, due to the fact that the moral reward structure is ultimately motived from pro-sociality, the machine takes this as its priority.

Though some requirements seem harder to satisfy (particularly relating to whether the machine has a concept of 'self' etc.), it is worth noting that this analysis just contains an initial look at whether the model above *might* satisfy the requirements set by the specification in the previous chapter. Obviously a more comprehensive assessment would need to be carried out, involving psychologically informed tests once the robot was actually created. What we have is a model that, at a preliminary glance, seems to satisfy the requirements listed in Chap. 6.

Reference

Piaget, J. (1936). *Origins of intelligence in the child*. Routledge & Kegan Paul.

Chapter 8
The Ethics of Machine Ethics

Abstract It's important to consider the ethical implications associated with this new field of study. Though Machine Ethics considers the ethics relating to how a machine might make ethical decisions, the system in of itself, has ethical ramifications. This chapter aims to consider seven of the main ethical issues. First; whether artificial moral agents are *safe* is considered, by discussing some previous arguments against Machine Ethics and whether bestowing machines with morals puts humanity at even more risk. Then, Machine Patiency or *Robot Rights* is discussed—*if robots have their own agency, should they have rights*? Then, there is consideration about whether such robots are sustainable, before broader ethical implications (an ethical impact assessment for moral machines) are outlined. Finally, new security implications associated with the approach to building moral machines is discussed, before the idea of *robot psychology* is introduced.

Keywords Machine ethics ethics · AI ethics · Ethics · Robot ethics · Robot psychology · Robot rights · Sustainability · AI safety · Impact · Robot security

We have discussed what 'morality' and 'machine' mean in a general sense, why we might need moral machines, we have considered an assurance-based framework from which to create and assess the machines and hypothesised a model for creating a moral machine inspired by a caregiver-child relationship. However, we have yet to consider the broader implications of this new kind of technology. For instance, if moral machines were to be successfully created and introduced into society, what wider considerations would need to be made—*what are the ethical issues relating to this new Machine Ethics?*

The aim of this chapter is to begin to consider some of these broader issues by introducing *new topics* that may well need to be explored in more depth if we are to have moral machines in society and the world. Some of the questions explored include, whether we need to ascribe some kind of moral status to these machines (the robot moral patiency debate), whether these machines warrant a new type of psychology to understand their behaviour (robot psychology), and what implications this new type of technology will have to humans and the world, i.e., what does it mean to have artificial agents with this cognitive capacity? There are also wider ethical questions pertaining to whether it is sustainable (from an environmental perspective) to have robots trained in this way, whether such a design is *private and secure* and whether the fears outlined against Moral Machines are allayed by this new approach, which will be considered first.

8.1 Better with Moral Machines

One of the key concerns for researchers working within Machine Ethics, is that despite best intentions, in creating moral machines, we will be doing more harm than good. As already outlined, as Winfield (2018b) articulates, the fear is that by creating a machine with the capacity to make moral decisions, we are increasing the chance of individuals making machines that make purely immoral decisions by making the job easier to achieve. The argument follows that in engineering machines to *be good* we are increasing the chance that machines are engineered to *be bad*, because the *good* can easily be switched to *bad*.

Conceptually, it seems that this argument assumes that there is some distinct faculty in the machine that ensures the machine always selects the morally good action in a situation. However, if the machine is not being controlled as to what constitutes 'good' and 'bad' behaviour in the first place, this faculty doesn't exist.

For example, in the case of Vanderelst and Winfield's Simulation architecture for morality (as detailed in Chap. 3), the machine's decision to carry out a good act over a bad one is governed by its 'Evaluation Module'. Essentially, an action only passes through this module—and goes on to be enacted—if it is one that is assessed as good. We can see how the existence of such a module might be problematic according to the argument above since it might be very easy to reverse the engineering in this module so that only the opposite holds true. This could pose a dangerous situation if, for instance, we had a medical robot making life-critical decisions, and the system were maliciously hacked. However, this problem is based upon the assumption that an ethical governance module (i.e., something that permits or restricts behaviour in a behaviour-constraining way) is the only way to create a moral machine.

What has been demonstrated over the previous chapters is a way to cultivate moral growth in a machine *without constraining its behaviour*. So, the issue stated dissolves away, because we are not creating any control mechanism for the machine's behaviour in any way that are easy to reverse. We are instead enabling the machine to make its own moral decisions through a secondary level feedback. There is no cognitive 'switch' to turn on or off, once the machine has been trained using the moral feedback layer, it should be secure in its objective in as much as a typically trained Reinforcement Learning-agent, once trained, cannot be made to develop a different model for how to optimise achievement of its task. Furthermore, according to the assurance framework, we would not be placing the machine into any kind of 'risky' situation until it had been sufficiently assessed to be—from a cognitive perspective—to have sufficient moral maturity anyway.

It might be argued that by introducing this secondary feedback module, that we are increasing the chance of 'bad' agents materialising because we can simply train the agent to believe that *rogue behaviours* are the acceptable form of behaviour, and therefore the machine learns *all the wrong ways* to, for instance, clean a house. Rather than having a robot that cleans the house in the right way, it always cleans the house by places pets into bins. It might be worth noting that if only *wrong* feedback were given to the machine, that eventually inconsistencies in feedback might lead to the machine becoming ultimately confused rather than inherently bad. We also have to take account of the fact that my 'good' might be different to another individuals 'good' (take, for example, differences in cultural moral beliefs), so the machines will be trained according to the cultural moral preferences of the teacher. This doesn't mean it will automatically accept the cultural moral preferences. Just because someone's metacommentary on how to carry out a task is different to my own, does not make this any less *right* and it will in fact be at the onus of the groups of responsible adults to either train their agents individually, or come to some consensus on the right way to do things.

This means that rather than making it easier to create 'bad' machines, the approach above will diversify the moral eco-system, meaning different cultural opinions will be catered for rather than argued against and debated. Creating moral machines in this way seems to pose no additional risk because all we are doing in finding a way to capture a variety of moral opinions, in creating a model to capture diversity in moral opinion, rather than constraining a machine's behaviour to operate according to one moral norm.

8.2 Machines with Rights

If machines are bestowed with the ability to make moral decisions, does this mean that they should be held accountable in the same way that humans are? Should machines be held accountable to the law, government policies, and importantly, should machines have rights in the same way that humans do? We have already discussed the idea that as children become moral morally mature, they are granted the ability to make more significant moral decisions, including voting and joining the army. If machines are proven to be morally mature in the same way, does this grant them rights in the same way that a voter has rights, should there be such things as *robot rights*?

There are significant discussions to be held in this area, with the introduction of machines with the ability to tell between right and wrong. According to many, because machines are ultimately mechanical devices, they do not warrant the same rights as humans and living creatures do, but such arguments can be regarded as dubious, particularly when we consider human-machine hybrid carnations (so-called 'cyborgs' in science fiction).

For instance, suppose an individual had a mechanical heart fitted to keep them alive, would this grant them a lower moral status because part of their body was made from a mechanical device? Many individuals have prosthetic limbs fitted to facilitate movement, but it seems obvious that we shouldn't grant them a lower degree of rights for this reason. As a result, the mechanical device argument against moral rights for machines seems problematic, especially as we consider machines that are capable of making decisions with a similar level of moral import as humans do.

Such academic issues have led many to fear the idea of creating machines with morals in the first place, because though the idea of bestowing artificial agents with the ability to decide between right and wrong does not seem scary in itself, intuitively, bestowing machines with equal rights as humans, does.

It is not obvious that machines should have the same degree of rights as humans at all, if they are so granted rights, or whether their rights should just be limited to having the right to make decisions with moral import (if the assurance permits), but there needs to be significant discussion in this area with the introduction of the new technology.

It would be unfair to discuss the topic of robot rights without mentioning that some academics have already considered the idea of whether artificial intelligence and robots (as a different type of technology) are deserving of rights. According to Coeckelbergh (2010), the moral status of an artefact may be prone to change when we make socio-relational considerations; that is, when we consider the role, the artefact plays in relation to society. The arguments following suggest that as robots have a social role in society, we ought to consider them as potentially moral subjects. Similar positions are presented by Gunkel (2018), and Gellers (2020) who considers what legal factors need to be thought of if we have agents with this new moral status ascription. Both topics—whether robots should have rights, and the legal framework to govern them—are topics that will require exploring in more depth as machines become more human in nature, including acquiring moral decision-making abilities.

8.3 Sustainability and Robots

It has been argued by many that there are issues when it comes to the Sustainability of robotic systems, particularly AI-embedded robotic systems which require a significant amount of energy for their training. As described in a paper I wrote on Sustainability and AI (Raper et al., 2022), many AI systems use up to 78,000 pounds of Carbon Dioxide when they are trained, which is the equivalence of three cars in their lifetime of use. If we are to substitute human decision-making with AI-trained systems, then are we only contributing to an already polluted world and causing problems to exacerbate issues associated with climate change rather than resolve them? Is there some way to design, develop and deploy AI-trained systems so that they either have minimal impact to the environment or reduce the effects that are already apparent?

For instance, although AI systems are often offered as a solution to solve climate challenges, such as weather prediction and smart energy consumption, the energy expended in creating these systems often outweighs the benefits, meaning that the solutions are in effect redundant or only make the situation worse. One immediate strategy to deal with this issue is to ensure that a thorough systems analysis is conducted prior to the design of a new AI system. For instance, the following questions need to be asked: will the benefits gained from using the new AI system outweigh the associated costs? Overall, from a sustainability perspective, what will be the environmental impact of the new system? If the answer to first question is 'no', and the environmental impact will be negative, then alternative methods should be considered as a solution to the sustainability problem.

As outlined in Raper et al. (2022), there can also be overarching management strategies to encourage more sustainable development generally. AI software developers can be incentivised (either through games or otherwise), to create systems that are energy friendly—receiving rewards for 'the most sustainable software solution'. For instance, it is possible to develop software programs with fewer lines of code, which require less energy to run. Rewards could drive competition between developers for the most 'efficient' code.

In terms of a thorough systems analysis related to the technological design, there is work to be done in understanding *the problem* that the new AI system is going to solve. Is the problem financially incentivized, or is it a social problem that will benefit the world, and will the AI solution actually serve as the best solution to the problem, or are more energy efficient solutions available? In the case of designing moral machines, there is significantly more work to be done in understanding in which situations we need agents with moral decisions. This book has provided an argument as to *why we need* machines with morals, but the specific case studies need exploring further, with a full portfolio provided. Sustainability considerations need to be made at each stage of the analysis.

When we start to consider physically engineering machines with moral agency (i.e., creating robots that can decide between right and wrong), there are other sustainability issues that need to be considered. If new materials are going to be created to make the robots, where will this come from? Will it use recycled resources, and what will happen if the machine/robot reaches its end-of-life, and the materials are no longer required? Is there some way to ensure that all the materials are repurposed, or can robots be made that do not have an end-of-life? What sustainable energy methods can be used to power or supply energy to the robots so that they do not use large amounts of carbon-exhaustive methods?

8.4 Moral Machines Impact Assessment

Integrating moral machines into case studies that require moral decision-making is bound to have some social implications in terms of affecting humans and having sociological impact that is not obvious to measure. For instance, if we have machines with the ability to make moral decisions, how will this affect our own moral decision-making?

There is significant agreement (and previous discussion) that moral decision-making is not something that should be outsourced, that is, we should not be using machines to make moral decisions on our behalf, but if they are not being used to replace human moral decision-making, what are they being used for? If we begin to consider such machines as having moral status in of themselves, does that mean we should start to consider the moral opinions of such machines as much as we do humans? If we did, would they represent legitimate moral opinions, or would they essentially pollute our human moral judgements?

If we take Coeckelbergh's (2010) socio-relational approach to assessing technology, then moral machines will have an impact on our lives. Inasmuch as we (as 'responsible adults') influence the decision-making system of the moral machines, the machine will have some bearing on our own moral decision-making. It's not impossible to imagine a future where humans and robots debate, for instance, who the next political party leader should be, whether 16 is an appropriate age to be permitted to vote, and whether euthanasia should be made legal. Inasmuch as we are the architects of the machine's cognitive moral agency, once introduced into society, the machines will be the architects of ours. Machine's already influence, for example, which television streaming programme we should watch next or song we should listen to through its AI-powered recommendation engines, but machines with moral agency will add an additional dimension to this influence that will need to be assessed through sociological impact assessments or otherwise.

As well as the sociological impact of this new type of technology, there is bound to also be some psychological impact felt by humans in terms of what relationships with robots in the future may look like, and what relationships we want to have with robots. Does the extra 'human-like' dimension warrant relationships with machines of a more romantic nature, and is this in itself morally permissible?

There is also the issue pertaining to ensuring that such machines do not deceive humans into believing that they are human. Though it is the essence of The Turing Test to assess the intelligence of an artificial agent based upon whether it can successfully deceive a human, to avoid a dystopia where we cannot tell what is real and is not, we need to be conscious to either make the distinction between machine and human still obvious, or to have some mechanism to distinguish between the two. As robotics is still not sufficiently progressed to having created mechanical devices that *look exactly* like humans, giving the *appearance* of a human is not yet a problem. However, remotely this is not the case, since some software AI systems, such as ChatGPT (c) are sophisticated enough to give the illusion of interacting with a human, and we also currently have the issue of *Deepfakes*—that is, videos that *look* like the real thing. Not only safety mechanisms are needed to mitigate against this risk, but there need to be laws and governance put in place to ensure developers of such systems put such safety mechanisms into place.

Similarly, as machines gain moral agency, we need to consider what implications this will have in terms of making the machines more human-like and manage this accordingly.

8.5 Secure by Design

It is not obvious whether additional levels of security to preserve privacy and confidentiality will need to be added to machines cultivated with moral agency, however it is worth considering *what these machines will know* and *how easily accessible this information will be* in respect of ensuring secure and safe design of moral robots. For instance, if trained on moral judgements from a responsible adult, does this mean that the machine will have access to a new bank of insights about the person training it—i.e. their wants, desires, and own motivations in life. If so, it is possible that this information could be maliciously hacked and used for nefarious purposes. This might not be a problem if the robot is being used solely by the individual training it, since only that individual will have access to the robot anyway, but if the robots are to be trained then disseminated more widely, how can such information be protected? Figure 8.1 gives an example of an awkward situation that might unfold if a robot were to disclose personal information in the wrong place.

Fig. 8.1 A robot disclosing unwanted information in a social situation

Secure by design is an approach suggesting that software architectures be designed with preservation of individual privacy and consent in mind from the start (similar to ethical by design). If this is the case, then there is some scoping to be done to assess what the security *risks* of this type of technology are, and it needs to be carried out as part of the overall agent design process, prior to implementation. For example, if such an agent has the ability to speak, can security measures be put in place so that it doesn't inappropriately disclose personal information when operating in a social setting?

8.6 Robot Psychology

The final broader (ethically related) issue that needs considering as machines gain more agency generally, and more cognitive capacities, such as the ability to tell between right and wrong, is our need to understand the machines *own psychology*.

This is distinct from the cognitive understanding of the machines, or the understanding of how to engineer the machines to have certain cognitive functions and requires understanding how the machines functions from a psychological perspective. For instance, the following questions might drive this field:

1. What are the machine's motivations?
2. Does the machine have inherent desires, and if so, what are they?
3. How does the machine process information from its environment to make decisions? [This question might seem like it has an obvious answer, i.e., using machine learning or reinforcement learning, but it is coming from a different angle. Rather than knowing that the machine uses a certain mathematical approach involving making predictions based upon historical data, we want to know how the machine represents concepts (such as tables and chairs), and how its processing is different to a human.]
4. What are the machine's cognitive abilities?

Not only are these questions linked to the assurance model described in earlier chapters, in that by answering them we are answering some of the questions posed in the test criteria, but as different types of agents, it is important to understand how they are different, and the same, to humans, from a scientific perspective.

The idea of a robot psychologist is not new and is in fact mentioned in Asimov's stories. In 'I-robot' we are introduced to Susan Calvin, a 'Robo-psychologist' whose job it is to understand how the robots *think*. Similarly, in Philip K Dick's (1968) 'Do Androids Dream of Electric Sheep', popularised by the *Bladerunner* film franchise, we are presented with an individual (the 'bounty hunter') who must carry out psychological tests to distinguish between 'androids' and 'humans'. Though purely science fiction until now, if we are to truly embrace robots and moral machines (in an arguably ethical way) robot psychology is a legitimate field that needs to be introduced and thoroughly explored.

References

Coeckelbergh, M. (2010). Robot rights? Towards a social-relational justification of moral consideration. *Ethics and Information Technology, 12*, 209–221.

Dick, P. K. (1996. [1968]). *Do androids dream of electric sheep?* Ballantine Books.

Gellers, J. C. (2020). *Rights for robots: Artificial intelligence, animal and environmental law (edition 1)*. Routledge.

Gunkel, D. J. (2018). The other question: Can and should robots have rights? *Ethics and Information Technology, 20*, 87–99.

Raper, R., Boeddinghaus, J., Coeckelbergh, M., Gross, W., Campigotto, P., & Lincoln, C. N. (2022). Sustainability budgets: A practical management and governance method for achieving goal 13 of the sustainable development goals for AI development. *Sustainability, 14*(7), 4019.

Chapter 9
Summary and Next Steps…

Abstract This chapter provides a summary of the rest of the book and then traces directions for future work in this area, first, by describing where we need to go to progress the general field of Machine Ethics, and then by giving a list of future work areas (*what we need to do*). A closing remark takes inspiration from a quote from Alan Turing.

Keywords Machine ethics · Summary · Conclusion · Future of robotics · Moral cognition · Future directions · Artificial intelligence

In under 70 pages, this book has introduced the topic of moral machines, it has defined concepts such as 'moral', 'machine' and more technical terms such as 'Artificial Intelligence' and 'Reinforcement Learning'. It has given a quick overview of *Machine Ethics* to date, that is, it has introduced Machine Ethics as an academic discipline, and provided a summary of the key debates within the area. The book has introduced the *Moral Cognitive Requirements Framework*—an approach that takes the Machine Ethics problem as one that needs a cognitive solution, rather than a behavioural one—and has provided a draft specification for what cognitive features a moral agent should have. Finally, a draft technical *model for morality* has been proposed that, based upon a meta commentary from a responsible adult—sets the key thesis of this book, that *robots should be raised to be good*. Finally, broader ethical considerations and associated topics have been introduced.

In this chapter, I quickly summarise the key lessons learned from discussion in this book, and stipulate where we need to go and what we need to do based upon these lessons.

9.1 What We Learned

Importantly, one of the first things we learned, is that morality has a rich and complex history and that it is not just something that can be 'solved' over the course of a few years. If we consider the philosophers, Kant, Aristotle, and Bentham who tried to somehow formalise morality, that is prescribe what it *means to be good*, each theory has its own problems. Where Kantianism fails on account of being too vague, Bentham's Utilitarianism fails because it is too specific. Aristotle was to some extent able to provide an account of the 'good person', but its vagueness means that it doesn't really provide direction for how to *be good*. In terms of defining 'the good' it seems that this could be a fruitless endeavour, or at least fruitless in terms of having a specific enough definition to be programmable into a machine. However, what the theories provide us with, are some understanding of the *types of features* that moral decision-making, from a cognitive perspective, constitutes. For instance, Virtue Ethics emphasises that 'good' behaviour should be habitual, part of our personal nature. From this we can extract the fact that morality is more than just one cognitive faculty but entwined in our human character and personality. If we are to create a machine with moral agency, this means that morality should be a fundamental part of its own character too.

We have learned that due to its complexity in nature, morality is not something that can be 'extracted' and placed into a machine. For instance, we can't create an algorithm for 'good' and just programme a machine to behave in this way. There are several reasons why this would not be possible: (1) because 'good' is variable among cultures and individuals, (2) because the 'good' changes (progresses) over time and (3) as per Virtue Ethics, just because something *behaves* in a good way, does not make it an inherently good thing. Instead, what we need to do is *cultivate moral growth*. This can only be done by allowing the machine to develop its *own perception of the right thing to do*, supported by our moral guidance.

If we train machines in this way, wider ethical and societal issues begin to materialise. First, we now have questions of whether these machines are deserving of moral rights; whether they should be bound by the same sorts of laws that humans are, and if not, whether new laws or governance mechanisms need inventing for these agents. The key priority at play here is *safety*. We need machines that are aligned to our best interests, but governance procedures and regulation to manage these new agents. We also need to manage the development of these machines by ensuring that they are not placed into making decisions with the *potential to harm*, until we are assured that their decisions are always going to be moral in nature. This links to the wider AI Safety debate, which concerns existential risks associated with sophisticated artificial intelligence. Ensuring that adequate moral maturity has been reached for the role undertaken, is essential here.

With new technology comes exciting new fields of scientific enquiry and exploration. Exciting new fields include *Robot Psychology* along with *a greater understanding of how morality develops in humans*. By learning how we can create machines with moral agency, we are enabling greater understanding and appreciation of the human mind and of what makes us who we are.

9.2 Where We Need to Go

This book should be viewed as a new problem; a new way of construing morality and the pursuit to create machines with morality. As such, we have two new fields of study that require significant further exploration.

Regarding the study of morality, we need to understand from a neurodevelopmental social perspective, *actually how children/humans develop morally*. Although Kohlberg's observations about different moral stages is a useful metric to measure growth, there is little work on assessing, cognitively, (a) why some individuals might develop maturity more than others and (b) what the mechanisms for development are. An approach to tackling this field, might involve looking into the reasons why morality does *not* develop in the first instance—potentially looking at individuals demonstrating psychopathic tendencies—or why moral development doesn't always work. The driving questions in this new 'Morality Studies' discipline should be, what are the mechanisms that facilitate moral growth, and why do they work? It has been suggested that fictions (such as parables and fairytales (see Currie (2005)) might play a role in the moral development of an individual. If so, why is this the case and can the process be replicated?

Regarding further research into Machine Ethics, viewing the problem as *a cognitive one* (rather than one where we want to replicate just moral behaviour) means we need to take the lessons from the new morality studies and learn to computationally replicate these. That is, rather than replicating *moral decisions,* we need to replicate the mechanism by which learning takes place. This should be viewed as a problem in Cognitive Robotics as much as a technical problem of Artificial Intelligence. In essence, how can we enhance the moral learning architecture?

9.3 What We Need to Do

There are four main key steps that need to be taken to develop our understanding in the area of Morality Studies, facilitate the growth of the new pursuit to create artificial moral agents, and manage the new technology that will be created through research into this area.

1. **We need to better understand the human moral cognitive faculty**: as mentioned above, this is so we can better understand, from a scientific perspective, essentially, how moral decisions are made. To satisfy this objective will require neuroscientific mapping of moral decisions along with brain studies to better understand not only what areas of the brain are involved in making a moral decision, but how the processing takes place.
2. **We need a more sophisticated information transfer model for moral knowledge acquisition:** from a logical perspective, what does moral knowledge constitute? If I state that I think killing is wrong, what does this mean? Semantics is the study of how we acquire meaning from language and logic. If we were to

carry out more research into moral semantics, we might better be able to model learning in the machine, which can only understand on the basis of logic.
3. **We need to explore what assurance means more thoroughly in the context of AI Safety:** this will involve exploring the motivations for moral machines and the context in which we need them. The assurance framework provided is just the first stage of research into this area, but we need to understand in more detail what test criteria would really give us this assurance.
4. **We need to investigate and develop a new regulatory framework surrounding the introduction of moral machines:** before moral machines are fully introduced into society, and before autonomous AI decision-making more generally is introduced into areas with significant implications, we need to ensure that they are safe. This will include developing the assurance framework mentioned above, but also tests for moral maturity and governance to ensure that developers are creating such autonomous machines in the right way.

9.4 Closing Remark

Alan Turing not only famously introduced us to the idea of thinking machines but was the founder of modern-day computing, inventing the notion of a digital computer to solve a mathematical puzzle known as *The Halting Problem*. Where we can define The Halting Problem as the question of whether a question is decidable or not (that is, whether it has an answer), Turing proposed that computers would never be able to solve undecidable questions. What constitutes as 'good' seems to be an undecidable question, to the extent that we have never been able to solve the question of what it means to be good. Therefore, we should not be trying to create machines that can solve this problem too. Solving the problem of morality is not about finding a solution to what good behaviour constitutes, but it is about a never-ending exploration into *how we ought to live*. Therefore, the best we can hope for is a machine that joins in with us on this pursuit of never-ending enquiry. One of the exciting things about morality and humanity as a whole is that we can keep learning about it forever. In the words of Turing (1950):

> We can only see a short distance ahead, but we can see plenty there that needs to be done.

Reference

Currie, G. (2005). Imagination and make-believe. In *The Routledge companion to aesthetics* (pp. 355–366). Routledge.

Glossary

AI Alignment   The field concerned with looking at how we might align Artificial Intelligence systems with human values.
Agency   Having the capacity to act to produce a particular effect.
Agent   A being with the capacity to act.
Agile **[methodology]**   An approach to management of projects that utilises the division of tasks into small iterative work phases.
Artificial Intelligence (AI)   The pursuit to recreate human intelligence.
Artificial General Intelligence (AGI)   A generalised model of human cognition that has the ability to carry out multiple tasks and functions.
AI Ethics   The discipline concerned with the ethical implications associated with Artificial Intelligence.
Algorithm   A set of rules or processes for problem-solving by a computer.
Architecture   The considered structure of something.
Assurance   Confidence that a particular set of events or behaviours that will take place.
Automated   Operated by automatic equipment.
Autonomous Vehicle   A vehicle that can operate on its own accord, without human intervention.
Bias   Prejudice towards one individual or group, that is considered to be unfair.
Brain   A soft tissue organ, sat within the skull, that acts as the centre of the nervous system.
Caregiver   An adult responsible as the key person looking after a child.
The Categorical Imperative   From Kant's thesis, a moral obligation that holds in all circumstances and is not dependent upon personal perspective.
Chat GPT   A piece of AI technology, utilising Large Language Models, that allows users to receive responses to prompts such as questions and documents.
Cognitive   Related to the mental processes involved in thought or reasoning.
Cognitive Science   The interdisciplinary field of study concerned with understanding the processes behind our thoughts.

Cognitive Robotics A sub-field of robotics concerned with understanding how we might bestow robots with intelligent behaviour.
Collaborative Robots (co-bots) A type of robot intended for human-robot-interaction, designed to work in close proximity to a human.
Computer An electronic device used for storing and processing information.
Computer Science The discipline concerned with understanding computers.
Conscience A person's guide to moral behaviour, their sense of 'right-' and 'wrong-'ness.
Consequentialism The theory that an action is judged as good (or bad) by the outcome of its consequences.
Conventional Level Part of Kohlberg's theory of moral development, this represents the second level, at which an individual begins to discern between societal norms and the concepts of 'right' and wrong'.
Cyborg A creature that is part human, part machine.
Deepfake A video which has been altered (often using AI) to represent a person or event that is not the case.
Democratise The process of making a thing accessible to everyone.
Deontological Ethics The name given to Kant's Ethical theory, the philosophical ethics concerned with duties, governed by The Categorical Imperative.
Developmental Psychology A sub-field of psychology concerning the growth or change of a person throughout their life.
Economics The discipline concerned with the production, distribution and consumption of goods and services.
Ecosystem A community of organisms and their environment.
Emotion A feeling derived from a person's mood or circumstances, that is distinct from rational thinking.
Ethical Representative of the right way to live.
Ethical by Design The doctrine that when we design technologies, we do so with ethics in mind.
Ethical Dilemma Famously introduced by Philipa Foot, with the introduction of The Trolley Problem, it describes a moral situation where there is no obvious right choice.
Ethics A sub-field of philosophy, concerned with the study of how we ought to live.
Ergon Taken from Aristotle's philosophy, the underlying task or function driving a creature's behaviour.
Eudaimonia From Aristotle's philosophy, the ultimate 'good' state of a human or being.
Explainability The idea that a concept or piece of technology can be explained to a relevant audience.
Fairnesss Treatment without prejudice or discrimination.
Flourishing Taken from Aristotle's philosophy: if something is serving its key purpose (it's 'ergon').
Framework Supporting structure to a system, concept or text.
Freedom The ability to act without restraint (physical or psychological).

Generative AI Artificial Intelligence that can generate either pictures, text or audio based upon an input prompt.
The Halting Problem A mathematical problem concerning whether a computer program will halt when given an input program.
Harm Injury inflicted by a person or artefact, which can manifest itself in many ways, including physically, psychologically or economically.
Human Rights Fundamental rights that exist simply because we are humans, regardless of nationality, gender, religious faith etc.
Humanity The collection of all beings that would be deemed humans.
Humanoid A machine that is designed and built to look or act like a human.
Industrial Robot A robotic system used in manufacturing, using consisting of a programmable arm with several degrees of movement.
Justice Treatment or behaviour that is fair and reasonable.
Large Language Model A type of Artificial Intelligence that analyses large quantities of text to detect patterns and understand and generate language.
Laws of Robotics Taken from Isaac Asimov's fictional stories, these prescribe a series of four (include the zeroth law) rules for robots to follow to ensure their behaviour is safe for humans.
Logic Reasoning that can be validated.
Machine A device created with several mechanical parts to carry out a dedicated function or task.
Machine Ethics A branch of computer science aimed at understanding how we might create machines with the ability to tell between right and wrong.
Machine Learning A statistical technique and branch of AI that makes predictions based upon input data.
Meta- A self-referential quality, something that refers to itself.
ML-Ops A branch of IT management that gives processes for the development of Machine Learning systems.
Mobile Robot A robot that is capable of moving around its environment (is mobile).
Model A representation of something that is usually smaller or simpler than the original thing.
Moral The principles for right and wrong behaviour.
Moral Agency The ability to make one's own moral decisions.
Moral Patiency The state of being eligible for moral consideration.
Moral Status How an agent ought to be treated based upon moral consideration.
Neural Network A machine learning approach designed to mimic the function and structure of the human brain.
Operating Software Computer software that manages a computer's hardware and resources.
Personality Combination of unique characteristics of a person that makes up who they are.
Philosophy The discipline concerned with the study of the nature of knowledge, reality and existence.
Policy A set of ideas or plans used as the basis for making decisions.

Post-Conventional Level Taken from Kohlberg's moral development theory, and represents the third stage, in which individuals become more understanding of individual liberty and human rights.

Pre-Conventional Level Taken from Kohlberg's moral development theory, and represents the first stage in which individuals take a more selfish stance towards morality.

Prima Facie A Latin expression, meaning 'at first impression', used to denote a theory or idea that is accepted until proven otherwise.

Principle A rule or belief governing behaviour.

Privacy The state of being unobserved by other people.

Psychology The discipline which considers the scientific study of the mind.

Relational The way in which two or more items are connected.

Reinforcement Learning (RL) A type of Machine Learning, where the agent optimises behaviour based upon expected rewards.

Robotics An interdisciplinary branch of computer science concerned with the design and development of machines with the ability to sense and react to their environment.

Robot A machine controlled by a computer that automatically reacts to its environment.

Secure by Design A principle dictating the design of technical systems so that security is considered at the outset.

Semantics A branch of linguistics concerned with the meaning of language.

The Simulation Theory of Cognition A theory of mind, that assert that we ascribe other's mental states through simulation.

Social Robots A type of autonomous robot that interacts and communicates with human users.

Sociology The discipline concerned with how society operates and the processes that underpin it.

Specification A detailed description of how something should be done or made.

Stakeholder Someone with an interest (a stake) in a new product or design.

State of Affairs A description of how things are involving a thing and its circumstances.

System A set of things that work together as a complex whole.

Superintelligent [AI] An agent with intelligence and cognitive capacities far surpassing that of a human.

Supervised Machine Learning A type of Machine Learning whereby the predicted outcome is guided or supervised with labelled data sets.

Sustainability The ability to be maintained at a certain level.

Systems Engineering A type of engineering concerned with analysing and understanding complex systems.

The Trolley Problem A well-known ethical dilemma, where the observer must decide whether an intervention to divert a trolley to save several people, sacrificing another individual, is morally permissible. There are variations of this problem involving different people and different interventions.

Theory of Mind A theory in psychology regarding the ability to ascribe mental states to another individual.
Transparency Operating in such a way that the actions are easy to see.
The Turing Test Often referred to as *the true test for artificial intelligence*, it prescribes a scenario whereby an individual must identify whether an artificial agent is a machine or human.
The Universal Maxim Given as the solution to Kant's Categorical Imperative—'act by any maxim that you would want it universalised'.
Utilitarianism A type of consequentialism, whereby the moral worth of a decision is determined by the utility of its outcomes—this could be 'happiness'.
Vice Defined by Kant as an immoral behaviour.
Virtue Defined by Kant as a moral behaviour.
Virtue Ethics The ethical theory put forth by Aristotle which asserts that to be moral we must act as the virtuous person would.

Index

A
Advisor, 21, 23, 68, 70
Agency, 3, 15, 35, 49, 64, 65
Agent(s), 4, 24–28, 30, 31, 37, 39, 42, 45, 46, 50, 51, 55–61, 65, 67, 74–80, 83–85
Agile (methodology), 49
AI alignment, 3, 20, 30–32, 35
AI ethics, 14, 15, 20, 21, 37
Algorithm(s), 16, 41, 84
Align(ing)(ed), 3, 30, 31, 45, 48, 59, 66, 71, 84
Allen, C., 15, 16, 22, 27, 28, 32
Analysis, 8, 24, 48, 51, 60, 66, 72, 77
Anderson, M., 21, 23, 25, 29
Anderson, S.L., 29
Architecture, 24, 45, 50, 71, 74, 85
Aristotle, 8, 25, 58, 84
Artificial agent, 4, 31, 78
Artificial General Intelligence (AGI), 30
Artificial Intelligence (AI), 3, 13, 20, 35, 49, 58, 76, 83
Artificial Intelligence (film), 2
Asimov, I., 1, 2, 21, 80
Assurance, 4, 42, 45–51, 60, 65–67, 70–74, 76, 80, 86
Authority, 5, 9
Automate(d), 3, 16, 23
Automatic, 37, 47
Autonomous, 3, 9, 12–16, 22, 27, 28, 35, 37, 39–42, 44, 55, 60, 61, 63, 68, 71, 86
Autonomous vehicle, 12, 20, 27, 68
Autonomy, 13, 15, 16, 22, 27, 35, 37, 39–41, 58

B
Bad, 7, 8, 56, 74, 75
Behaviour(s), 1, 7, 22, 44, 54, 66, 73, 84
Bentham, J., 10, 25, 84
Bias, 14, 21, 22, 29, 32
Big Hero 6 (film), 2
'Bottom-up' approach, 22
Brain, 85
Bryson, J., 26, 39

C
Caregiver, 47, 66–68, 71, 73
(The) Categorical Imperative, 9, 10, 25
ChatGPT, 3, 13, 30, 36, 61, 78
Child, 4, 9, 11, 37, 47, 66–71, 73
Cognitive, 5, 13, 21, 24, 40, 45, 50, 51, 54, 55, 61–63, 65, 66, 68, 74, 78, 80, 83, 85
Cognitive requirements, 4, 48–51, 54, 55, 57, 61, 62, 65, 83
Cognitive robotics, 85
Cognitive science, 13
Collaborative robots (co-bots), 13
Computer, 2, 3, 11, 12, 17, 30, 86
Computer games, 2
Computer science, 3
Conscience, 9, 72
Consequentialism, 10, 11, 24, 25, 54–56
Constrain, 44, 46, 47, 68, 71
Control, 12, 13, 23, 24, 30, 50, 55, 56, 66, 67, 74
Conventional level, 9, 28

Critical, 4
Crowdsource(d), 23, 24, 29, 32
Cyborg, 75

D
Data, 14, 22, 30, 31, 80
Decision-making, 3, 5, 9–12, 14, 15, 21, 22, 24, 26–28, 37, 41–43, 45, 54, 55, 61, 67, 68, 76–78, 84, 86
Deepfake, 78
Delphi, 31, 32
Democratis(e)(ing), 46
Deontological ethics, 54, 59–60
Design, 4, 10, 15, 21, 48–50, 53, 60, 61, 66, 70, 72, 74, 76, 77, 79
Designing, 4, 10, 65, 77
Detroit: Become Human (computer game), 2
Developmental nature, 10
Developmental psychology, 4
Developmental stage(s), 9, 66
Duty, 10, 25, 59

E
Economics, 3
Ecosystem, 4
Embody, 25
Emotion(s), 61
Equality, 9
Ergon, 58
Ethical, 4, 9, 20, 36, 44, 54, 66, 73, 83
Ethical by design, 21, 79
Ethical decision machines, 20–22
Ethical dilemma(s), 11, 20, 21, 23, 29, 30, 37
Ethical principles, 9, 20, 21
Ethics, 3, 7, 19, 35, 45, 54, 66, 73, 83
Eudaimonia, 25, 58
Evaluation, 29, 74
Explainability, 21
Explicit ethical AI, 20

F
Fairness, 9, 21, 28
Film(s), 2, 27, 41, 46, 80
Flourishing, 25, 26, 58
Foot, Philipa, 11
Framework, 4, 19, 22, 43–51, 53, 60, 65, 66, 73, 74, 76, 83, 86
Free, 8, 36, 47, 58, 59, 64, 68, 72
Freedom, 46, 47, 59, 65
Full moral agency, 15

G
Generative AI, 61
Good, 2–5, 7–10, 12, 17, 19, 25–27, 36, 45–47, 54, 56, 57, 61, 66–69, 74, 75, 83, 84, 86
Grow, 5, 47
Growth, 4, 28, 58, 66, 67, 70, 74, 84, 85

H
Habit, 8, 12, 25
(The) Halting Problem, 86
Happiness, 10, 11
Harm, 1, 2, 20, 26, 40, 41, 44, 46, 51, 56, 66, 74, 84
Healthcare, 21, 22
Hedonistic, 9
Hinton, G., 30
Historical, 4, 80
Human(s), 1, 12, 20, 36, 43, 54, 66, 75, 84
Humanity, 2, 20, 30, 31, 86
Humanoid(s), 1, 36, 46
Human right(s), 9
Humans (TV Programme), 2
Husbands, P., 12, 13, 15
Hybrid, 21, 22, 32, 75

I
Immoral, 27, 40, 42, 44, 46–47, 74
Impact, 1, 5, 27, 41, 42, 56, 63, 67, 76–78
Implicit ethical AI, 20, 40
Industrial robot, 16
Infra-red, 12, 13
Interdisciplinary, 3, 13, 53
I-Robot (Fiction), 2, 80

J
Judgement(s), 11, 27, 29, 31, 32, 40–45, 78, 79
Justice, 9, 10, 22

K
Kant, I., 10, 25, 59, 84
Knowledge, 3, 14, 25, 27, 55, 56, 61, 71, 85
Kohlberg, L., 8–10, 28, 85

L
Large language models, 30
Law, 1, 2, 7, 9, 10, 25, 28, 61, 68, 75, 78, 84

Laws of Robotics, 1
Learner, 68
Logic, 36, 38, 86
Love, 2, 5

M
Machine(s), 1, 7, 19, 35, 43, 53, 65, 73, 83
Machine Ethics, 3–5, 7, 15, 16, 19–32, 35, 37, 39–41, 45, 46, 49, 66, 73–80, 83, 85
Machine learning (ML), 14, 49, 80
Machine Learning–Operations (ML-Ops), 49
Machines with morals, 1–5, 19, 26–28, 31, 44, 45, 76–78, 84, 85
Maxim, 10
Mental state, 24, 54
McCarthy, J., 13
Mechanical device, 12, 16, 17, 47, 75, 76, 78
MedEthEx, 21–24, 31
Medical, 21–24, 68, 74
Methodology, 4, 19, 48, 49, 51, 53, 62
Meta, 68, 70, 83
Metrics, 4, 8, 28, 49
Mimicking, 28
Mind, 10, 13, 21, 23, 28, 30, 35, 42, 49, 50, 54, 56, 60, 79, 84
Mobile robot(s), 13
Model, 4, 5, 14, 23, 25, 27, 30, 43, 45, 49, 50, 61, 66, 67, 69–75, 80, 83, 85, 86
Moor, J., 20, 40
Moral(s), 1, 7, 19, 35, 43, 53, 65, 73, 83
Moral agency, 15, 16, 28, 35, 38–42, 44, 46–47, 50, 51, 54–59, 62–66, 68, 70, 74, 77–79, 84
Moral behaviour, 1, 10, 22, 25, 45, 50, 54, 59, 68, 70, 71, 85
Moral decision(s), 3, 9, 12, 22, 23, 26, 37, 39, 45, 54, 55, 68, 74, 75, 77, 78, 85
Moral development, 5, 9, 28, 47, 68, 85
Morality, 2–5, 7–10, 12, 15, 16, 19, 21–30, 35, 44, 45, 50–51, 53–74, 83–86
Morality studies, 85
Moral judgements, 27, 31, 78, 79
Moral machine(s), 1–5, 7, 10, 12, 15–17, 19–21, 23, 24, 26–29, 31, 32, 35–45, 48–51, 53, 57, 58, 60, 66, 67, 70–75, 77–78, 80, 83, 86
(The) Moral Machine Experiment, 23, 29, 31
Moral nature, 10
Moral patiency, 4, 73
Moral rights, 5, 76, 84
Moral status, 73, 75, 76, 78
Motivated, 56, 63, 64, 71

Motivation, 4, 26, 35, 43, 44, 47, 56, 71, 79, 80, 86

N
Nao (robot), 13
Nature, 4, 9, 10, 12, 16, 23, 26, 27, 31, 45, 47, 49, 54, 59, 61, 65, 70, 76, 78, 84
Neural networks, 14
Nussbaum, M., 58

O
Operating software, 12

P
Personality, 25, 57, 63, 72, 84
Personhood, 26
Philosophy, 3, 20, 45, 54, 56–59
Philosopher(s), 8, 10, 11, 15, 58, 84
Philosophical, 3, 4, 24, 25, 45, 54, 55, 57
Piaget, J., 66
Policy, 8, 75
Post-conventional level, 9, 28
Practice, 4, 68
Pre-conventional level, 9, 28
Predict, 24, 25, 30, 47, 63
Prima facie, 21
Principle(s), 2, 9, 10, 20–25, 28
Privacy, 79
Private, 74
Processes, 21, 29, 40, 42, 46, 49, 50, 53–55, 61, 65, 66, 68, 69, 79, 80, 85
Processing, 54, 80, 85
Psychology, 3, 54, 73
Punishment, 9

Q
Quality, 8, 30, 49, 72

R
Raised to be good, 47, 83
Raising a robot, 3–5, 66
Rawls, J., 21
Reciprocity, 9, 10
Reflective equilibrium (approach), 21
Regulation, 21, 84
Relational, 59
Relationship, 26, 45–47, 54, 59, 65–68, 71, 73, 78

Relativity, 3, 31
Reinforcement learning (RL), 14, 31, 46, 58, 69–71, 80, 83
Requirement(s), 4, 16, 27, 35, 37–39, 41, 42, 44, 48–51, 53–60, 62–66, 70–72
Research, 3, 4, 19, 20, 32, 35, 45, 53, 85, 86
Responsibility, 41, 49, 66, 67
Responsible, 41, 49, 66–68, 71, 72, 75, 78, 79, 83
Responsible engineering, 20, 21
Right, 9–11, 13, 15, 16, 20, 29, 45, 50, 66–70, 75–77, 80, 86
Right thing to do, 9, 44, 84
Risk, 3, 11, 15, 21–24, 26, 27, 30, 31, 39, 40, 42, 51, 56, 70, 74, 75, 78, 79, 81
Robot and Frank (film), 2, 27
Robotics, 3, 14, 61, 76, 78
Robot(s), 1–4, 12–17, 24, 27, 35–41, 46, 62, 66–70, 72, 74–80, 83
Robot psychology, 73, 80, 84
Robot rights, 75, 76
Robot vacuum cleaner, 13
Roomba (c), 15, 37, 40
Rogue AI, 58
Rule(s), 1, 2, 7, 9, 10, 17, 21, 22, 28
Runaround, 1

S
Safety, 2, 3, 5, 15, 20, 30, 37, 39–41, 44, 66, 78, 84, 86
Scientific, 27, 45, 54, 80, 85
Scientific enquiry, 5, 84
Secure, 74, 79
Secure by design, 79
Self, 56, 57, 63, 71, 72
Semantics, 85, 86
Sensors, 13, 36, 37, 40, 54
Simulation, 24, 74
(The) Simulation Theory of Cognition, 24
Social robots, 13, 35
Social situation, 4, 79
Social standards, 7
Society, 5, 9, 16, 24, 36, 73, 76, 78, 86
Societal impact, 1, 24
Sociology, 3
Specification, 4, 48–51, 53, 60–64, 66, 72, 83
Stakeholder(s), 51
Standard(s), 7–9, 17, 22, 36, 43, 66
State of affairs, 56, 57, 63, 71

System, 3, 14–16, 20–23, 29, 31, 32, 41–43, 48, 49, 54, 58, 60, 61, 68, 71, 74, 76–78
Systematic, 5
Subjective nature of morality, 12
Superintelligent (AI), 30, 31
Supervised machine learning, 14
Sustainable, 74, 77
Sustainability, 5, 76–77
Systems engineering, 48

T
Tay, 31
(The) Ten Commandments, 9
Tesla, 36, 37
Test(s)(ing), 4, 28–30, 44, 48–51, 53, 60–66, 68, 70, 72, 80, 86
The Moral Machine Experiment, 23, 29, 31
Theory of Mind, 24
The Trolley Problem, 11, 12, 23, 25, 37, 46, 55, 61
The Wizard of Oz (film), 2
Thinking, 2, 25, 28, 47, 61, 66, 86
'Top-down' approach, 22
Transparency, 21
Trust, 27, 44–47, 50, 65, 66
Trustworthy, 65
TV Programmes, 2, 41
Turing, A., 20, 28, 86
(The) Turing Test, 28, 29, 78
Turtle-bot, 13–15
Twitter, 31

U
Unbiased, 21
Unethical, 46, 68
Unitree Robodog, 13
Universal, 9, 10, 25
(The) Universal Maxim, 10, 25, 59
Unpredictable, 26, 44, 46
Utilitarianism, 10, 25, 84

V
Vice(s), 8, 25
Vicious, 8
Virtue(s), 8, 25, 57
Virtue ethics, 12, 25, 57–58, 84
Virtuous, 8, 25, 57, 58

W
Wallach, W., 15, 16, 22, 27, 28, 32
(The) Waterfall Method (TWM), 48–50, 53, 60
Well-being, 36, 40, 41, 66

Winfield, A., 24, 27, 74
Wrong, 4, 9, 15, 16, 20, 22, 24, 41, 45, 66, 75–77, 79, 80, 85

Printed in the United States
by Baker & Taylor Publisher Services